# ROBOT DYNAMICS ALGORITHMS

## THE KLUWER INTERNATIONAL SERIES IN ENGINEERING AND COMPUTER SCIENCE

ROBOTICS: VISION, MANIPULATION AND SENSORS

*Consulting Editor*

TAKEO KANADE

**Other books in the series:**

*Robotic Grasping and Fine Manipulation, M. Cutkosky.*
ISBN 0-89838-200-9

*Shadows and Silhouettes in Computer Vision, S. Shafer.*
ISBN 0-89838-167-3

*Perceptual Organization and Visual Recognition, D. Lowe*
ISBN 0-89838-172-X

*Three-Dimensional Machine Vision, T. Kanade, Editor.*
ISBN 0-89838-188-6

# ROBOT DYNAMICS ALGORITHMS

by

**Roy Featherstone**
Edinburgh University

Springer Science+Business Media, LLC

**Library of Congress Cataloging-in-Publication Data**

Featherstone, Roy.
Robot dynamics algorithms.

(The Kluwer international series in engineering and computer science ; SECS 22. Robotics)
Based on the author's thesis—Edinburgh University, 1984.
Bibliography: p.
Includes index.
1. Robots—Dynamics. 2. Dynamics, Rigid.
3. Recursive functions. I. Title. II. Series:
Kluwer international series in engineering and computer science ; SECS 22. III. Series: Kluwer international series in engineering and computer science. Robotics.
TJ211.4.F43 1987 629.8'92 87-3260
ISBN 978-1-4757-6437-6 ISBN 978-0-387-74315-8 (eBook)
DOI 10.1007/978-0-387-74315-8

# Table of Contents

# Preface

The purpose of this book is to present computationally efficient algorithms for calculating the dynamics of robot mechanisms represented as systems of rigid bodies. The efficiency is achieved by the use of recursive formulations of the equations of motion, i.e. formulations in which the equations of motion are expressed implicitly in terms of recurrence relations between the quantities describing the system. The use of recursive formulations in dynamics is fairly new, so the principles of their operation and reasons for their efficiency are explained.

Three main algorithms are described: the recursive Newton-Euler formulation for inverse dynamics (the calculation of the forces given the accelerations), and the composite-rigid-body and articulated-body methods for forward dynamics (the calculation of the accelerations given the forces). These algorithms are initially described in terms of an un-branched, open-loop kinematic chain -- a typical serial robot mechanism. This is done to keep the descriptions of the algorithms simple, and is in line with descriptions appearing in the literature. Once the basic algorithms have been introduced, the restrictions on the mechanism are lifted and the algorithms are extended to cope with kinematic trees and loops, and general constraints at the joints. The problem of simulating the effect of contact between a robot and its environment is also considered. Some consideration is given to the details and practical problems of implementing these algorithms on a computer.

The algorithms are presented using a six-dimensional vector notation called spatial notation, which is similar to the use of screw coordinates to represent vector and tensor quantities. This notation greatly simplifies the analysis of rigid-body dynamics by reducing the number and size of the equations involved. It simplifies the process of formulating the algorithms and allows the finished algorithms to be expressed clearly and concisely.

The reader is assumed to have a knowledge of vectorial mechanics. A knowledge of Lagrangian mechanics would be helpful, but is not necessary. No prior knowledge of robot dynamics, or the dynamics of any other kind of rigid-body system, is required.

## Acknowledgments

My thanks to John Hallam and Robert Gray, who read through the draft version of this book and made many useful comments; and also to Magdalena Müller, who took on the arduous task of formatting the equations.

# ROBOT DYNAMICS ALGORITHMS

NOTE: Where Greek letters have been used to denote vectors or matrices, it was intended that they should appear in a bold typeface to distinguish them from scalars. Unfortunately this has not happened, and it is possible that some confusion may arise where the same letter has been used to denote both a vector and a scalar.

# Chapter 1
# Introduction

## 1.1. Scope and Contents

This book is concerned with mathematical formulations of the dynamics of robot mechanisms which produce efficient algorithms when implemented on a computer. A robot mechanism is assumed to be a system of rigid bodies connected by ideal joints and powered by ideal force generators. In the sense that a robot mechanism is a rigid-body system, the algorithms described here for robot dynamics are also algorithms for the dynamics of certain classes of rigid-body system.

The efficiency derives from the use of recursive formulations of the equations of motion; i.e., formulations in which the equations of motion of the system as a whole are expressed implicitly in terms of recurrence relations between quantities describing properties of the system. The mathematical formulations are derived with the aid of a 6-dimensional vector notation, called spatial notation, which greatly facilitates the analysis of rigid-body dynamics by reducing the size and number of equations involved. This notation is similar to the use of screw coordinates to represent vector and tensor quantities.

The book is organised into two parts. Part one deals with the spatial notation and its application to spatial rigid-body dynamics, and part two deals with the robot dynamics algorithms.

Three main algorithms are described: the recursive Newton-Euler method for inverse dynamics, and the composite-rigid-body and articulated-body methods for forward dynamics. Inverse dynamics is the problem of finding the joint forces required to produce a given acceleration, and finds application in robot control; whereas forward dynamics is the problem of finding the instantaneous acceleration of a robot given the applied forces, and finds application in simulation.

The algorithms are described initially for an un-branched, open-loop kinematic chain, which is typical of a serial robot mechanism. The descriptions of the algorithms are kept as simple as possible. Once the basic algorithms have been introduced, modifications are described which enable them to cope with branched kinematic chains and a greater variety of constraints at the joints. The details of implementing the algorithms on a computer are then considered; and finally, algorithms are described for calculating the forward dynamics of robot mechanisms with kinematic loops, and for simulating the effect of contact between a robot and its environment.

This book is concerned only with presenting exact, general-purpose, numerical algorithms. Approximate algorithms, special-purpose algorithms (i.e., special to a particular manipulator geometry), and techniques for manual or automatic derivation and simplification of the symbolic equations of motion are not considered. Applications of the dynamics algorithms are mentioned, since the algorithms are designed to be of practical use, but no particular application is described in any detail.

## 1.2. The Robot Dynamics Problem

In principle, solving the robot dynamics problem presents no difficulty -- a robot mechanism is just a system of rigid bodies, and the equations of motion of such systems have been known for a long time. The real problem is the practical one of finding formulations of robot dynamics that lead to efficient computation algorithms.

In its simplest form, the equation of motion for a robot mechanism can be expressed as a vector differential equation

$$\ddot{\mathbf{q}} = \phi(\,\mathbf{q}, \dot{\mathbf{q}}, \tau\,)\,,$$

where $\mathbf{q}$ is the vector of generalised (position) coordinates, $\dot{\mathbf{q}}$ and $\ddot{\mathbf{q}}$ are its derivatives with respect to time, and $\tau$ is the input force vector. To solve this equation for $\mathbf{q}$ requires procedures for evaluating $\phi$ and for performing numerical integration. For rigid-body systems of the complexity of a robot mechanism the main problem is the evaluation of $\phi$.

With simple rigid-body systems, the most obvious way to solve the dynamics is to obtain expressions for the system accelerations directly in terms of the system parameters and variables; i.e., express $\phi$ explicitly in terms of $\mathbf{q}$, $\mathbf{q}$, $\tau$ and the system parameters. The result is a set of symbolic, scalar equations of motion, one for each degree of freedom, and computing the acceleration is simply a matter of evaluating these equations with given numeric values.

Unfortunately, this approach does not work well for larger systems, since the number of terms in the symbolic equations grows very rapidly with the number of bodies. As the equations get bigger, they also become very repetitive; i.e., they contain large numbers of similar terms, each differing from the others by only a few factors. The evaluation of the equations involves a great deal of duplicated, and hence unnecessary, calculation.

What is needed is a structured evaluation. Commonly occurring sub-expressions must be identified, and a calculation scheme must be adopted which calculates the values of these expressions, uses the results to evaluate more expressions, and so on until the answer is reached.

All dynamics algorithms for complex rigid-body systems evaluate the equations of motion in stages. The results of each stage are a set of values which will be used in subsequent stages of the calculation. The equations of motion are a property of the system itself, but the intermediate steps are a property of the algorithm. So the differences between dynamics algorithms lie in the structure they impose on the calculation through the intermediates they specify.

The execution time for different algorithms varies enormously, but the differences are accounted for by the amount of unnecessary calculation involved in evaluating the equations of motion via a prescribed set of intermediates. The key to efficient dynamics calculation is to find a set of common sub-expressions which is effective in eliminating most of the repetition inherent in the equations of motion, and to specify these as the intermediate results to be calculated by the algorithm.

A technique that has been found to be particularly effective in producing efficient dynamics algorithms is that of formulating the dynamics recursively in terms of recurrence relations [10]. The task is broken down into a number of partially ordered stages, each of which is solved by the application of a formula to each link in turn. Where appropriate, the formula defines the quantity of interest for the link in question in terms of the values of that quantity for one or more of the link's immediate neighbours, in which case it is known as a recurrence relation. The practice of using previously calculated results in subsequent applications of recurrence relations ensures that the intermediate results are frequently-occurring common sub-expressions of the overall dynamics equations (but not necessarily the optimum set).

The basic dynamics algorithms described in this book are all formulated in this manner. A practical example of how recurrence relations work is given in Chapter 4.

## 1.3. Spatial Notation

A particle has three degrees of freedom, so its velocity, acceleration, etc. can each be described by a 3-dimensional vector, and it has one vector equation of motion. A rigid body, on the other hand, has six degrees of freedom, yet rigid-body dynamics is conventionally expressed using 3-dimensional vectors. So the velocity, acceleration, etc. of a rigid body are each represented by a pair of vectors (one linear and one angular) and there are two equations of motion. The reason for this is that linear and angular quantities are considered to be physically different quantities; but this does not preclude their amalgamation into 6-dimensional vectors as a matter of algebraic and notational convenience.

The spatial notation is based on the use of 6-dimensional vectors, called spatial vectors, to represent quantities like the total velocity or acceleration of a rigid body. The use of spatial notation results in a considerable reduction in the number of quantities and in the number and size of equations required to express, solve, or express the solution to a problem in spatial rigid-body dynamics. This is particularly so for problems involving the manipulation of rigid-body inertias. Spatial vectors are represented by $6 \times 1$ matrices of components, and spatial tensors by $6 \times 6$ matrices. The operations of spatial algebra are implemented using the operations of standard matrix arithmetic, with the exception that a different transpose operator is used. Spatial notation is similar to the use of screw coordinates to represent vector and tensor quantities.

The spatial notation is described in part one of this book, and is used in part two to derive the dynamics algorithms. Part one also serves as an introduction to the use of spatial notation in the analysis of spatial rigid-body dynamics, and many of the techniques and concepts used in part two are described there. A small number of examples are included, which illustrate or expand on some of the points made in the text.

### 1.4. Readers' Guide

This book is organised into two parts. Part one describes the spatial vector algebra and comprises Chapters 2 and 3, while part two describes the robot dynamics algorithms and comprises Chapters 4 to 9.

In part one, Chapter 2 introduces the basic concept of a spatial vector and deals with the kinematic aspects of the spatial vector algebra, while Chapter 3 is concerned with the representation of inertia, the equations of motion and spatial vector analysis. A summary of spatial algebra is included at the end of Chapter 3. Part one also serves as an introduction to the use of spatial notation in the analysis of spatial rigid-body dynamics. Readers acquainted with the use of screws or motors in kinematics will probably find most of the material in Chapter 2 familiar but most of the material in Chapter 3 unfamiliar.

In part two, all of the algorithms are described using spatial notation, so a proper understanding of part two can only be obtained by reading part one first. However it is possible to get the gist of part two by taking the spatial formulae on trust and appealing to the superficial similarity between spatial formulae and their counterparts in conventional vector algebra.

Chapters 4 to 6 describe the three robot dynamics algorithms in their simplest form. Chapter 4 describes the recursive Newton-Euler method for inverse dynamics, Chapter 5 the composite-rigid-body method for forward dynamics, and Chapter 6 the articulated-body method for forward dynamics. Chapter 4 also defines the system model used in the descriptions of the algorithms, and explains by way of an example how recurrence relations work. Chapters 4 and 5 begin with some introductory material on robot dynamics, which describes briefly the historical development, the approaches taken, etc., and mentions the applications of the algorithms. The important points concerning articulated-body inertias (which were introduced in Chapter 3) are reiterated in Chapter 6.

Chapter 7 extends both the basic algorithms and the robot system model so that kinematic trees and more general constraints at the joints can be accommodated. Algorithms for robots with floating (i.e., kinematically unconstrained) bases, and hybrids between forward and inverse dynamics are also described. The descriptions are necessarily less detailed than those in Chapters 4 to 6, and wherever possible the extensions are described in terms of modifications to the basic dynamics algorithms.

Chapter 8 is concerned with implementation issues such as the choice of link coordinate systems and the computer representation of spatial quantities. The kinematic preliminaries and other details that were ignored in earlier chapters are described here. Also, a comparison of the computational requirements of the basic dynamics algorithms is presented. This chapter will be of interest to anyone wishing to implement the algorithms described in this book.

Chapter 9 deals with the dynamics of robot mechanisms containing kinematic loops, and with the effects of contact and impact between a robot and objects in its environment. Kinematic loops are treated as constraint forces which are applied to a tree mechanism. The treatment of contacts is similar, but is made more complicated by the possibility of contacts being broken.

# Chapter 2
# Spatial Kinematics

## 2.1. Introduction

A spatial vector is a 6-dimensional vector which can be used to represent the combined linear and angular components of the physical quantities involved in rigid-body dynamics. The usefulness of spatial vectors and related quantities lies in their power to simplify the description and analysis of rigid-body systems: a single spatial vector can hold the information which would otherwise need at least two 3-dimensional vectors, and a single spatial equation can replace two or more 3-dimensional vector equations. The result is that rigid body systems can be described by fewer equations, relating fewer quantities, and the equations are usually shorter than their 3-dimensional counterparts.

This chapter describes the representation of physical vector quantities as spatial vectors; the relationship between spatial vectors, line vectors and free vectors; spatial coordinate systems and coordinate transformations; and the differentiation of spatial vectors. Finally, spatial vectors are compared with the closely related concepts of screws and motors. The representation of inertia and the equations of motion for a rigid body will be dealt with in Chapter 3.

For a proper understanding of spatial algebra, the reader must be prepared to accept that quantities of a physically different nature, like linear and angular velocity, can be combined as components of a single entity; and it is important to regard spatial quantities as objects in their own right rather than as conglomerations of more familiar objects.

## 2.2. Spatial Velocity

Let us consider how the velocity of a rigid body can be expressed as a spatial vector. A rigid body has six degrees of freedom -- three translational and three rotational -- so the location, displacement, or velocity of a rigid body can be described completely by six numbers: three describing the translational aspect and three the rotational aspect. Conventionally, the velocity of a rigid body would be described by a linear and an angular velocity vector. The question is, how should they be assembled to form a spatial vector?

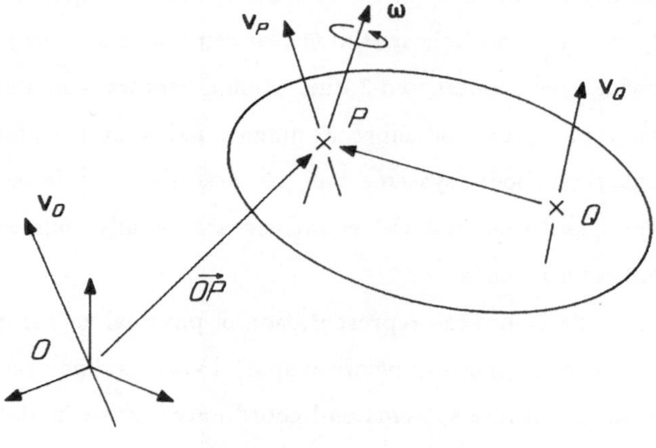


**Figure 2-1:** Rigid Body Velocity

The instantaneous velocity of a rigid body may be described by the linear velocity $\mathbf{v}_P$ of some point $P$ in the rigid body and its angular velocity $\omega$ (see Figure 2-1). $\omega$ applies to the body as a whole, and is independent of the choice of $P$, but $\mathbf{v}_P$ applies only to the point $P$. The instantaneous velocity $\mathbf{v}_Q$ of any other point $Q$ in the rigid body can be expressed in terms of $\omega$ and $\mathbf{v}_P$ as

$$\mathbf{v}_Q = \mathbf{v}_P + \vec{QP} \times \omega \, .$$

The instantaneous velocity of the rigid body is described completely by the pair of vectors ( $\omega$, $\mathbf{v}_P$) given the point $P$; and the body may be considered to be rotating with angular velocity $\omega$ about an axis passing through $P$ whilst simultaneously translating with linear velocity $\mathbf{v}_P$.

The choice of $P$ is arbitrary. It need not be physically within the rigid body, so long as it has the same motion as the body. Moreover, we need not choose the same point relative to the body from one instant to the next. So let us choose a point which is moving with the rigid body, but which is instantaneously coincident with the origin, and represent the velocity of the rigid body by the pair ( $\omega$, $\mathbf{v}_O$ ). $\mathbf{v}_O$ is the velocity of the point at the origin, and is given by

$$\mathbf{v}_O = \mathbf{v}_P + \vec{OP} \times \omega \,. \tag{2.1}$$

The body may now be considered to be rotating about an axis passing through the origin whilst simultaneously translating with linear velocity $\mathbf{v}_O$. The factor $\vec{OP} \times \omega$ added to the linear velocity at $P$ has had the effect of shifting the axis of rotation from passing through $P$ to passing through $O$.

The pair ( $\omega$, $\mathbf{v}_O$ ) defines the spatial velocity of the rigid body. We shall represent spatial vector quantities as 6-element column vectors and denote all spatial quantities with carets. Thus in spatial notation we say that $\hat{\mathbf{v}}$ is the spatial velocity of the rigid body, where $\hat{\mathbf{v}} = [\, \omega_x \, \omega_y \, \omega_z \, v_{Ox} \, v_{Oy} \, v_{Oz} \,]^T$ which will normally be abbreviated to $\hat{\mathbf{v}} = [\, \omega^T \, \mathbf{v}_O{}^T \,]^T$.

## 2.3. Line Vectors and Free Vectors

An 'ordinary' 3-dimensional vector has magnitude and direction, but a vector may also have positional properties, depending on the nature of the physical (or mathematical) entity it represents. In particular, two kinds of vector occur in rigid-body dynamics: line vectors and free vectors.

A line vector has magnitude and direction, and is constrained to lie in a definite line (i.e., a line in a particular position). Line vectors arise naturally in mechanics to represent quantities like angular velocity, where there is a definite axis of rotation, and forces acting on a rigid body, which have a definite line of action but may act at any point along that line.

A free vector has magnitude and direction only, and is used to represent quantities like linear velocity and couple which have no definite line or axis associated with them.

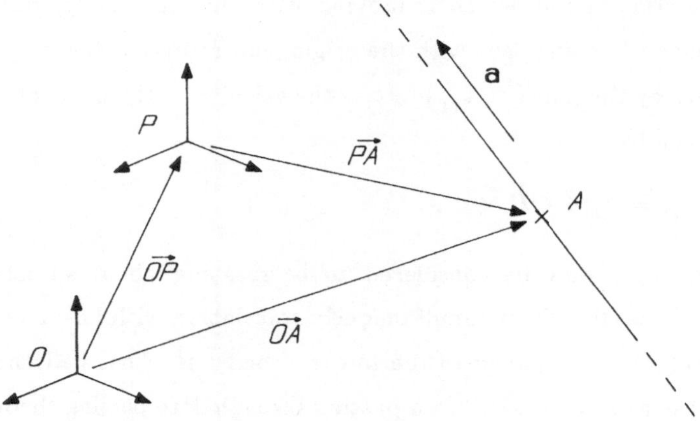


**Figure 2-2:** Definition of a line vector

A line vector may be represented as follows. Suppose a line vector has magnitude and direction given by the vector **a** and its line of action passes through the point $A$ (see Figure 2-2). Since the direction of the line is the same as that of the line vector, we may say that the vector has magnitude and a line of action. Its magnitude is that of **a**, and the equation of the line of action is given by

$$(\mathbf{r} - \vec{OA}) \times \mathbf{a} = \mathbf{0},$$

where **r** is the position of any point on the line. This equation is more normally stated in the form

$$\mathbf{r} \times \mathbf{a} = \mathbf{a}_O , \tag{2.2}$$

where $\mathbf{a}_O = \vec{OA} \times \mathbf{a}$. $\mathbf{a}$ and $\mathbf{a}_O$ are known as the Plücker vectors of the line and determine it completely. $\mathbf{a}$ is called the resultant vector and $\mathbf{a}_O$ the moment vector. Although six numbers are specified, only five are independent since $\mathbf{a}\cdot\mathbf{a}_O = 0$; and if the vectors represent just the line then one number is arbitrary since $\lambda\mathbf{a}$ and $\lambda\mathbf{a}_O$ represent the same line as $\mathbf{a}$ and $\mathbf{a}_O$, though not the same line vector. Line vectors are described in more detail in [11] and [28].

$\mathbf{a}_O$ depends on the choice of origin, for if the origin is moved from $O$ to $P$ then the Plücker vectors become $\mathbf{a}$ and $\mathbf{a}_P$ where $\mathbf{a}_P = \vec{PA} \times \mathbf{a}$. So we have

$$\mathbf{a}_O = \mathbf{a}_P + \vec{OP} \times \mathbf{a} , \tag{2.3}$$

which is the same shifting formula as for spatial velocity. The only difference between the line vector and the spatial velocity vector, $[\, \omega^T \, \mathbf{v}_O{}^T \,]^T$, is that it is not necessary for $\omega\cdot\mathbf{v}_O$ to be zero. A line vector can be expressed as a spatial vector $\hat{\mathbf{a}} = [\, \mathbf{a}^T \, \mathbf{a}_O{}^T \,]^T$; indeed any spatial vector where $\mathbf{a}\cdot\mathbf{a}_O = 0$ represents a line vector.

A free vector with magnitude and direction given by the vector $\mathbf{b}$ is of course adequately represented by $\mathbf{b}$. We can put free vectors on the same algebraic footing as line vectors by expressing them in the form $\hat{\mathbf{b}} = [\, \mathbf{0}^T \, \mathbf{b}^T \,]^T$, for such a quantity obeys the shifting formula of Equation (2.3) and yet is independent of the choice of origin.

### 2.3.1. Line and Free Vector Components of Spatial Vectors

A general spatial vector is the sum of a line vector and a free vector. There are an infinite number of ways in which a given spatial vector can be constructed from the sum of a line vector and a free vector; but if we constrain the line vector to pass through a given point then the line and free vector components of a spatial vector are determined uniquely. The decomposition of a spatial vector into its line and free vector components is particularly simple if the line vector is constrained to pass through the origin, in which case the line and free vector components of a spatial vector $\hat{\mathbf{a}} = [\, \mathbf{a}^T \, \mathbf{a}_O^{\,T} \,]^T$ are $[\, \mathbf{a}^T \, \mathbf{0}^T \,]^T$ and $[\, \mathbf{0}^T \, \mathbf{a}_O^{\,T} \,]^T$ respectively.

A general spatial vector is also equivalent to a screw vector, for we can express $\hat{\mathbf{a}} = [\, \mathbf{a}^T \, \mathbf{a}_O^{\,T} \,]^T$ uniquely as the sum of a line vector and a parallel free vector:

$$\hat{\mathbf{a}} = \begin{bmatrix} \mathbf{a} \\ \mathbf{a}'_O \end{bmatrix} + \begin{bmatrix} \mathbf{0} \\ \rho \mathbf{a} \end{bmatrix}$$

where $\mathbf{a}'_O = \mathbf{a}_O - \rho\, \mathbf{a}$ and $\rho = \frac{\mathbf{a} \cdot \mathbf{a}_O}{\mathbf{a} \cdot \mathbf{a}}$. Such a quantity is known as a screw vector in analogy with the motion of a nut on a screw thread -- it rotates about an axis and translates parallel to that axis. The line of action of $[\, \mathbf{a}^T \, \mathbf{a}'^{\,T}_O \,]^T$ is called the screw axis, and the parameter $\rho$ is called the pitch; both are invariants of the screw.

### 2.3.2. The Representation of Joint Axes

The essential feature of a joint is that it allows some degree of relative motion between the two bodies that it connects. In the case of a one-degree-of-freedom joint, the relative position of the two bodies is a function of a scalar called the joint variable, and we can express the relative velocity of the two bodies in terms of a joint axis and a scalar called the joint velocity (or rate), which is the derivative of the joint variable. The joint axis is a spatial vector which defines the direction and nature of motion allowed by the joint, and the relative velocity of the two bodies is obtained by multiplying the joint axis by the joint velocity.

A revolute (or turning) joint allows rotation about an axis fixed in both bodies. A revolute joint axis can be represented by a unit line vector, which describes both the position of the axis and the rotational nature of the motion (angular velocity being a line vector). The line vector must have unit magnitude in order that the unit joint velocity correspond to a unit relative angular velocity between the two bodies.

A prismatic (or sliding) joint allows pure translation between two bodies in a particular direction, and is characterised only by the permitted direction of motion; so a unit free vector is the appropriate representation for the joint axis.

A screw joint combines rotation about a definite axis with translation along it, so it can be represented by the sum of a line vector and a parallel free vector. The pitch of the screw determines the ratio of the magnitudes of the line and free vectors, and a unit screw axis is defined as one having a unit line vector. Further discussion of screw joints, and the representation of joints allowing more than one degree of motion freedom, is deferred until Chapter 7.

Once a joint axis is represented as a spatial vector, the type of motion allowed by the joint becomes irrelevant. There is no need to have separate equations to deal with each different joint type.

**Example 1**

Let $\hat{\mathbf{s}} = [\, \mathbf{s}^T \, \mathbf{s}_O{}^T \,]^T$ be the axis of a joint connecting bodies $b_1$ and $b_2$, and let $\dot{q}$ be the (scalar) joint velocity indicating the relative velocity of $b_2$ with respect to $b_1$, then the relative spatial velocity of $b_2$ with respect to $b_1$ is $\hat{\mathbf{s}}\,\dot{q} = [\, \mathbf{s}^T\dot{q} \;\; \mathbf{s}_O{}^T\dot{q} \,]^T$. If $b_1$ and $b_2$ have spatial velocities of $\hat{\mathbf{v}}_1$ and $\hat{\mathbf{v}}_2$ respectively, then

$$\hat{\mathbf{v}}_2 = \hat{\mathbf{v}}_1 + \hat{\mathbf{s}}\,\dot{q}\,.$$

This simple equation holds for any type of one-degree-of-freedom joint, and it expresses the relationship between both the linear and angular velocities of the two bodies.

**Example 2**

A robot has $n$+1 links numbered 0 . . $n$ and $n$ joints numbered 1 . . $n$. Link 0 is stationary, and link $i$−1 is connected to link $i$ by joint $i$. The axis of joint $i$ is $\hat{\mathbf{s}}_i$ and its joint velocity is $\dot{q}_i$ measured from link $i$−1 to link $i$. The relative velocity of link $i$ with respect to link $i$−1 is then $\hat{\mathbf{s}}_i \dot{q}_i$, and the absolute velocity of link $i$ is

$$\hat{\mathbf{v}}_i = \sum_{j=1}^{i} \hat{\mathbf{s}}_j \dot{q}_j .$$

In particular, the velocity of link $n$, the end effector, is $\hat{\mathbf{v}}_n = \sum_{j=1}^{n} \hat{\mathbf{s}}_j \dot{q}_j$. This can be expressed in matrix form as

$$\hat{\mathbf{v}} = [\hat{\mathbf{s}}_1 \ldots \hat{\mathbf{s}}_n] \begin{bmatrix} \dot{q}_1 \\ \cdot \\ \cdot \\ \cdot \\ \dot{q}_n \end{bmatrix} = \hat{\mathbf{J}}\,\dot{\mathbf{q}} .$$

The 6 × $n$ matrix, $\hat{\mathbf{J}}$ , composed of the axis vectors is the Jacobian of the transformation between the manipulator's joint space and the Cartesian velocity of the end effector expressed in the same coordinates as the $\hat{\mathbf{s}}_i$.

### 2.3.3. The Representation of Forces

A force acting on a rigid body has magnitude and a line of action. It may therefore be represented as a line vector. A spatial force with magnitude and direction given by **f** and acting along a line passing through a point $P$ is given by $\hat{\mathbf{f}} = [\,\mathbf{f}^T \ (\overrightarrow{OP} \times \mathbf{f})^T\,]^T$. The moment vector has the physical dimensions of a couple, and the rules for shifting the line of action of a force by adding a couple are the same as the corresponding rules for shifting line vectors. A pure couple has magnitude and direction only, and may therefore be represented as a free vector. A general spatial force is the sum of a pure force and a couple, and can be resolved into a force acting through the origin and a couple. If several spatial forces are acting on a single rigid body, then their resultant is a single spatial force which is simply the vector sum of the individual forces.

## 2.4. Spatial Vectors

A spatial vector is a member of a 6-dimensional vector space over the real numbers. To establish a correspondence between spatial vectors and the vectors of the 3-dimensional Euclidean space, certain spatial vectors are identified as line vectors and certain others as free vectors. To make the correspondence explicit, a Cartesian coordinate frame is placed in the 3-dimensional space and a *standard basis* is defined in the spatial vector space. This basis, denoted $\{\hat{\mathbf{e}}_i\}$, consists of six vectors $\hat{\mathbf{e}}_1 \,.\,.\, \hat{\mathbf{e}}_6$, where $\hat{\mathbf{e}}_1 \,.\,.\, \hat{\mathbf{e}}_3$ are unit line vectors along the $x$, $y$ and $z$ axes of the Cartesian coordinate frame and $\hat{\mathbf{e}}_4 \,.\,.\, \hat{\mathbf{e}}_6$ are unit free vectors parallel to the $x$, $y$ and $z$ axes of that frame. There is a unique standard basis for each Cartesian coordinate frame in 3-dimensional space. A general spatial vector $\hat{\mathbf{a}}$ may be expressed in terms of $\hat{\mathbf{e}}_1 \,.\,.\, \hat{\mathbf{e}}_6$ as

$$\hat{\mathbf{a}} = \sum_{i=1}^{6} a_i \hat{\mathbf{e}}_i \,, \tag{2.4}$$

where the scalars $a_i$ are the coordinates of $\hat{\mathbf{a}}$ in $\{\hat{\mathbf{e}}_i\}$. $\hat{\mathbf{a}}$ may be represented in $\{\hat{\mathbf{e}}_i\}$ by the $6 \times 1$ matrix of coordinates $[\, a_1 \; \ldots \; a_6 \,]^T$. All of the spatial vectors we have met so far have been expressed in terms of their coordinates in a standard basis.

To clarify the distinction between the spatial vector itself and its matrix representation, I will occasionally denote the matrix representation with a tilde, thus:

$$\widetilde{\mathbf{a}} = [\, a_1 \; \ldots \; a_6 \,]^T .$$

$\hat{a}$ is a primitive entity, but $\tilde{a}$ is a representation of $\hat{a}$ in a particular coordinate system. The expression $\hat{a} + \hat{b}$ is independent of any coordinate system, whereas $\tilde{a} + \tilde{b}$ is only meaningful if $\tilde{a}$ and $\tilde{b}$ are representations in the same coordinate system. Most of the time this distinction will be ignored and $\hat{a}$ will be used to denote both the vector and its matrix representation.

### 2.4.1. Operations on Spatial Vectors

For an expression involving spatial quantities to make physical sense there must be a physical interpretation for the mathematical operations in the expression as well as for the spatial quantities themselves. The interpretation of a mathematical operation depends on what operation it is and what its operands are. For the most part the restrictions are pretty obvious and much as you would expect, for example it does not make sense to add a velocity to a force, but there are a few cases where interpretations valid for ordinary vectors do not carry over to spatial vectors. (This is particularly so with the scalar product defined in Chapter 3.) This topic is dealt with rigorously by Woo and Freudenstein [77] and more intuitively by Ball [4], and will not be covered here. Instead, here is a list of the permissible interpretations of operations on those spatial quantities we have met so far.

- Addition is defined between spatial forces acting on the same rigid body, the sum being the resultant force.

- Addition is defined between spatial velocities representing the velocity of A with respect to B and B with respect to C, the sum being the velocity of A with respect to C.

- Multiplication of a spatial vector representing a joint axis by a scalar representing the joint velocity (i.e., the first time derivative of the joint variable) gives a spatial vector representing the relative velocity between the two links connected by the joint.

## 2.5. Rigid-Body Coordinate Transformations

Let us consider the transformations of spatial vector space which correspond to rigid-body transformations in 3-dimensional Euclidean space. A rigid-body transformation, or Euclidean displacement, is one which leaves the distance between any two points invariant. A rigid-body transformation in three dimensions consists of a rotation and a translation, each having three degrees of freedom. The image of a Cartesian coordinate frame under a rigid-body transformation is another Cartesian coordinate frame.

Let $\{\hat{\mathbf{e}}_i\}$ and $\{\hat{\mathbf{f}}_i\}$ be standard bases corresponding to Cartesian coordinate frames at $O$ and $P$ respectively, then the linear spatial mapping $\hat{\mathbf{Y}}: \hat{\mathbf{e}}_i \to \hat{\mathbf{f}}_i$ corresponds to the rigid-body transformation that maps $O$ coordinates to $P$ coordinates.[1] $\hat{\mathbf{Y}}$ will be referred to as a spatial rigid-body transformation.

Using tildes to denote matrix representations, suppose that a vector $\hat{\mathbf{a}}$ is represented as $\tilde{\mathbf{a}}_O = [\, a_{O1} \;\cdots\; a_{O6} \,]^T$ in $\{\hat{\mathbf{e}}_i\}$ and as $\tilde{\mathbf{a}}_P = [\, a_{P1} \;\cdots\; a_{P6} \,]^T$ in $\{\hat{\mathbf{f}}_i\}$, so

$$\hat{\mathbf{a}} = \sum a_{Oi}\, \hat{\mathbf{e}}_i = \sum a_{Pi}\, \hat{\mathbf{f}}_i \,. \tag{2.5}$$

The mapping $\tilde{\mathbf{X}} : \tilde{\mathbf{a}}_O \to \tilde{\mathbf{a}}_P$ is the spatial rigid-body *coordinate* transformation induced by $\hat{\mathbf{Y}}$ -- if we apply $\hat{\mathbf{Y}}$ to the vectors of a basis, then we must apply $\tilde{\mathbf{X}}$ to the coordinate matrix of a vector represented in that basis in order to obtain its representation in the new basis. $\tilde{\mathbf{X}}$ is a passive transformation in that it changes the representation of a spatial vector, as opposed to an active transformation, like $\hat{\mathbf{Y}}$, which maps a vector to a different vector (its image). $\tilde{\mathbf{X}}$ is linear, being the coordinate transformation induced by $\hat{\mathbf{Y}}$, so it may be expressed as a $6 \times 6$ matrix.

---

[1]It can be shown, using the one-to-one correspondence between standard bases and Cartesian coordinate frames, that a linear mapping in Euclidean 3-space must correspond to a linear mapping of spatial vector space.

The rest of this section is devoted to finding $\widetilde{\mathbf{X}}$ in terms of the rigid-body transformation that maps $O$ coordinates to $P$ coordinates.

To translate the origin of coordinates from $O$ to $P$ we use the shifting formula of Equation (2.3) (or Equation (2.1)), which gives

$$\hat{\mathbf{a}}_P = \begin{bmatrix} \mathbf{a} \\ \mathbf{a}_O + \vec{PO} \times \mathbf{a} \end{bmatrix}, \quad \text{where } \hat{\mathbf{a}}_O = \begin{bmatrix} \mathbf{a} \\ \mathbf{a}_O \end{bmatrix}. \tag{2.6}$$

This can be rearranged to express $\hat{\mathbf{a}}_P$ as the product of a $6 \times 6$ matrix with $\hat{\mathbf{a}}_O$ thus:

$$\hat{\mathbf{a}}_P = \begin{bmatrix} \mathbf{1} & \mathbf{0} \\ \vec{PO}\times & \mathbf{1} \end{bmatrix} \hat{\mathbf{a}}_O . \tag{2.7}$$

Here $\mathbf{0}$ is the $3 \times 3$ zero matrix, $\mathbf{1}$ is the $3 \times 3$ identity matrix, and $\vec{PO}\times$ is a $3 \times 3$ anti-symmetric matrix obtained by applying the cross operator, $\times$, to $\vec{PO}$. The cross operator is defined as follows:

$$\begin{bmatrix} x \\ y \\ z \end{bmatrix} \times = \begin{bmatrix} 0 & -z & y \\ z & 0 & -x \\ -y & x & 0 \end{bmatrix}. \tag{2.8}$$

The product of the matrix $\mathbf{a}\times$ with the vector $\mathbf{b}$ is equal to the vector cross product $\mathbf{a} \times \mathbf{b}$ (which I will write as $\mathbf{a}\times\,\mathbf{b}$ when I wish to emphasise that $\mathbf{a}\times$ is an object in its own right). The cross operator is a linear operator with the following properties:

$$\lambda(\mathbf{a}\times) = (\lambda\,\mathbf{a})\times, \qquad (\lambda \text{ is a scalar}) \tag{2.9}$$

$$\mathbf{a}\times + \mathbf{b}\times = (\mathbf{a} + \mathbf{b})\times, \tag{2.10}$$

$$(\mathbf{a}\times)^T = -\,\mathbf{a}\times, \tag{2.11}$$

$$(\mathbf{a}\times)\,\mathbf{b} = -(\mathbf{b}\times)\,\mathbf{a}, \tag{2.12}$$

$$(\mathbf{E}\,\mathbf{a})\times = \mathbf{E}\,\mathbf{a}\times\,\mathbf{E}^{-1}, \quad (\mathbf{E} \text{ is a } 3 \times 3 \text{ orthogonal matrix}) \tag{2.13}$$

$$(\mathbf{a} \times \mathbf{b})\times = \mathbf{a}\times\,\mathbf{b}\times - \mathbf{b}\times\,\mathbf{a}\times . \tag{2.14}$$

Unless indicated otherwise, expressions such as $\mathbf{a} \times \mathbf{b} \times \mathbf{c}$ are to be interpreted as $(\mathbf{a}\times)(\mathbf{b}\times)\mathbf{c}$, which is equivalent to the vector triple product $\mathbf{a} \times (\mathbf{b} \times \mathbf{c})$ but not to $(\mathbf{a} \times \mathbf{b}) \times \mathbf{c}$.

To effect a coordinate rotation, say from $O$ coordinates to $O'$, consider the spatial vector $\mathbf{a}$ to be the sum of a line vector through the origin and a free vector:

$$\hat{\mathbf{a}} = \begin{bmatrix} \mathbf{a} \\ \mathbf{0} \end{bmatrix} + \begin{bmatrix} \mathbf{0} \\ \mathbf{a}_O \end{bmatrix} .$$

The representation of $\hat{\mathbf{a}}$ in the rotated coordinate system is the sum of the rotated representations of the line vector and the free vector.

Let $\mathbf{E}$ be the $3 \times 3$ orthogonal rotation matrix which performs the coordinate rotation, then the representation of the line vector, which still acts through the origin, in the rotated coordinate system is $[(\mathbf{Ea})^T \mathbf{0}^T]^T$, and the representation of the free vector is $[\mathbf{0}^T (\mathbf{Ea}_O)^T]^T$. So if $\hat{\mathbf{a}}_O$ is a spatial vector represented in $O$ and $\hat{\mathbf{a}}_{O'}$ is the same vector in $O'$ then the coordinate transformation is

$$\hat{\mathbf{a}}_{O'} = \begin{bmatrix} \mathbf{E} & 0 \\ 0 & \mathbf{E} \end{bmatrix} \hat{\mathbf{a}}_O . \tag{2.15}$$

Spatial coordinate transformations will be given leading and following subscripts indicating the destination and source coordinate systems respectively. Thus ${}_O\hat{\mathbf{X}}_P$ is a spatial transformation which operates on a vector represented in coordinate system $P$ and produces a representation of the same vector in coordinate system $O$ according to the formula

$$\hat{\mathbf{v}}_O = {}_O\hat{\mathbf{X}}_P \hat{\mathbf{v}}_P . \tag{2.16}$$

(Note the pattern of subscripts.)

Spatial coordinate transformations may be combined by matrix multiplication, thus

$$ {}_O\hat{X}_P \, {}_P\hat{X}_Q = {}_O\hat{X}_Q \, . \tag{2.17} $$

The form of a general spatial coordinate transformation, which involves both rotation and translation, may be obtained by multiplying the transformation matrices of Equations (2.7 and 2.15). The spatial transformation corresponding to a shift of origin from $O$ to $P$ followed by a rotation about $P$ is

$$ {}_P\hat{X}_O = \begin{bmatrix} \mathbf{E} & 0 \\ 0 & \mathbf{E} \end{bmatrix} \begin{bmatrix} 1 & 0 \\ \mathbf{r}\times^T & 1 \end{bmatrix} = \begin{bmatrix} \mathbf{E} & 0 \\ \mathbf{E}\,\mathbf{r}\times^T & \mathbf{E} \end{bmatrix}, \tag{2.18} $$

where $\mathbf{r} = \overrightarrow{OP}$. The inverse transformation is

$$ {}_O\hat{X}_P = \begin{bmatrix} 1 & 0 \\ \mathbf{r}\times & 1 \end{bmatrix} \begin{bmatrix} \mathbf{E}^{-1} & 0 \\ 0 & \mathbf{E}^{-1} \end{bmatrix} = \begin{bmatrix} \mathbf{E}^{-1} & 0 \\ \mathbf{r}\times\mathbf{E}^{-1} & \mathbf{E}^{-1} \end{bmatrix}, \tag{2.19} $$

which is not the transpose of ${}_P\hat{X}_O$, so unlike $3 \times 3$ rotation matrices general spatial transformations are not orthogonal.

There are several ways of representing spatial displacements/coordinate transformations, most of which are described in [51]. From the computational point of view, the use of a $6 \times 6$ matrix of real numbers is one of the least efficient, both in terms of storage space and computation time. However, the machine implementation of an operation and its mathematical representation are largely independent, and Chapter 8 describes methods for achieving the desired result (i.e., a change of representation) on a computer which are far more efficient than using $6 \times 6$ matrices.

## 2.6. Differentiation in Moving Coordinates

The absolute time derivative of a spatial vector $\hat{\mathbf{a}}$ represented in the coordinate system $\{\hat{\mathbf{e}}_i\}$ is

$$\frac{d}{dt}\hat{\mathbf{a}} = \frac{d}{dt}(\sum_{i=1}^{6} a_i \hat{\mathbf{e}}_i)$$

$$= \sum_{i=1}^{6} ((\frac{d}{dt}a_i)\hat{\mathbf{e}}_i + a_i(\frac{d}{dt}\hat{\mathbf{e}}_i)) . \tag{2.20}$$

If the coordinate system is stationary then $\frac{d}{dt}\hat{\mathbf{e}}_i = \hat{\mathbf{0}}$ and the absolute derivative of $\hat{\mathbf{a}}$ is just the component-wise derivative. The problem to be discussed here is that of differentiating a spatial vector represented in a moving coordinate system.

The rigorous approach is to obtain an expression for the derivative of a moving vector which can be substituted for $\frac{d}{dt}\hat{\mathbf{e}}_i$ in Equation (2.20), but this approach is difficult to follow and will not be described here.[2] A simpler approach is to transform the vector to a stationary coordinate system, differentiate it there, and transform the derivative back to the moving coordinate system. This makes the (correct) assumption that the derivative of a spatial vector can be transformed like a spatial vector.

Let $P$ be a moving coordinate system, $O$ a stationary one, and ${}_O\hat{\mathbf{X}}_P$ and ${}_P\hat{\mathbf{X}}_O$ the transformations between the two. Using $\frac{d}{dt}$ to denote absolute differentiation and $\frac{d'}{dt}$ to denote apparent, or component-wise differentiation, the absolute derivative of a vector $\hat{\mathbf{a}}$ represented in $P$ coordinates is given by

$$\frac{d}{dt}\hat{\mathbf{a}} = {}_P\hat{\mathbf{X}}_O \frac{d'}{dt}({}_O\hat{\mathbf{X}}_P \hat{\mathbf{a}})$$

---

[2] The differentiation of moving screws and motors are described in [77] and [11], respectively. The differentiation of moving spatial vectors follows the same line of reasoning.

$$= \frac{d'}{dt}\hat{\mathbf{a}} + {}_P\hat{X}_O \left( \frac{d'}{dt}\, {}_O\hat{X}_P \right) \hat{\mathbf{a}} \, . \tag{2.21}$$

The problem is to evaluate ${}_P\hat{X}_O(\frac{d'}{dt}\, {}_O\hat{X}_P)$. This is done by considering the components of ${}_O\hat{X}_P$. If the transformation from $O$ to $P$ coordinates is achieved by a translation $\mathbf{r}$ followed by a rotation $\mathbf{E}^{-1}$ (a $3 \times 3$ orthogonal matrix), then

$${}_O\hat{X}_P = \begin{bmatrix} \mathbf{E} & 0 \\ \mathbf{r}\times \mathbf{E} & \mathbf{E} \end{bmatrix},$$

$$\frac{d'}{dt}\, {}_O\hat{X}_P = \begin{bmatrix} \frac{d'}{dt}\mathbf{E} & 0 \\ (\frac{d'}{dt}\mathbf{r})\times \mathbf{E} + \mathbf{r}\times \frac{d'}{dt}\mathbf{E} & \frac{d'}{dt}\mathbf{E} \end{bmatrix}. \tag{2.22}$$

Let $P$ coordinates have velocity $\hat{\mathbf{v}}$ expressed in $O$ coordinates as $\hat{\mathbf{v}}_O = [\, \omega^T \, \mathbf{v}_O{}^T \,]^T$. Now $\frac{d'}{dt}\mathbf{r} = \mathbf{v}_P = \mathbf{v}_O - \mathbf{r} \times \omega$ and $\frac{d'}{dt}\mathbf{E} = \omega \times \mathbf{E}$ (differentiation of an orthogonal matrix, e.g., see [7]), so

$$\frac{d'}{dt}\, {}_O\hat{X}_P = \begin{bmatrix} \omega\times \mathbf{E} & 0 \\ (\mathbf{v}_O - \mathbf{r}\times \omega)\times \mathbf{E} + \mathbf{r}\times (\omega\times \mathbf{E}) & \omega\times \mathbf{E} \end{bmatrix}$$

$$= \begin{bmatrix} \omega\times \mathbf{E} & 0 \\ \mathbf{v}_O\times \mathbf{E} + \omega\times \mathbf{r}\times \mathbf{E} & \omega\times \mathbf{E} \end{bmatrix} \quad \text{(by Equation (2.14 ))}$$

$$= \begin{bmatrix} \omega\times & 0 \\ \mathbf{v}_O\times & \omega\times \end{bmatrix} \begin{bmatrix} \mathbf{E} & 0 \\ \mathbf{r}\times \mathbf{E} & \mathbf{E} \end{bmatrix}$$

$$= \begin{bmatrix} \omega\times & 0 \\ \mathbf{v}_O\times & \omega\times \end{bmatrix} {}_O\hat{X}_P \, . \tag{2.23}$$

Introducing the spatial cross operator, $\hat{\times}$, defined by

$$\begin{bmatrix} \mathbf{a} \\ \mathbf{b} \end{bmatrix} \hat{\times} = \begin{bmatrix} \mathbf{a}\times & 0 \\ \mathbf{b}\times & \mathbf{a}\times \end{bmatrix}, \tag{2.24}$$

Equation (2.23) becomes

$$\frac{d'}{dt}\, {}_O\hat{\mathbf{X}}_P = \hat{\mathbf{v}}_O \hat{\times}\, {}_O\hat{\mathbf{X}}_P\,. \tag{2.25}$$

The spatial cross operator is the spatial analogue of the cross operator described earlier. It is linear and obeys rules equivalent to those given in Equations (2.9-2.14) but with spatial vectors and coordinate transformations in place of the ordinary vectors and $3 \times 3$ matrices.[3] The expression $\hat{\mathbf{a}} \hat{\times} \hat{\mathbf{b}}$ is the spatial equivalent of the motor product of motors [11], [17].

If we let $\hat{\mathbf{v}}_P = {}_P\hat{\mathbf{X}}_O\, \hat{\mathbf{v}}_O$ be the velocity of $P$ coordinates represented in $P$ coordinates then we get

$$\frac{d}{dt}\hat{\mathbf{a}} = \frac{d'}{dt}\hat{\mathbf{a}} + \hat{\mathbf{v}}_P \hat{\times} \hat{\mathbf{a}}\ ; \tag{2.26}$$

so the absolute derivative of a vector represented in a moving coordinate system is the sum of its apparent derivative and the cross product of the absolute velocity of the coordinate system with the vector being differentiated. This is closely analogous to the rule for differentiating an ordinary vector in a rotating coordinate system.

Equation (2.26) may be used to find the derivative of a moving vector as follows. Let the vector $\hat{\mathbf{a}}$, represented in stationary coordinates $O$, be moving with velocity $\hat{\mathbf{v}}$. Let $P$ be a coordinate system moving with velocity $\hat{\mathbf{v}}$, so $\hat{\mathbf{a}}$ is stationary in $P$, then

$$\begin{aligned} \frac{d}{dt}\hat{\mathbf{a}} &= {}_O\hat{\mathbf{X}}_P\left(\frac{d'}{dt}\hat{\mathbf{a}}_P + \hat{\mathbf{v}}_P \hat{\times} \hat{\mathbf{a}}_P\right) \\ &= {}_O\hat{\mathbf{X}}_P\left(\hat{\mathbf{v}}_P \hat{\times} \hat{\mathbf{a}}_P\right) \end{aligned}$$

---

[3]In the case of Equation (2.11) it is necessary to replace the transpose operator with the spatial transpose operator described in Chapter 3.

$$= \hat{\mathbf{v}} \times \hat{\mathbf{a}}\ . \tag{2.27}$$

If $O$ is a moving coordinate system and $\hat{\mathbf{v}}$ is the apparent velocity of $\hat{\mathbf{a}}$ in $O$, then Equation (2.27) gives the apparent derivative of $\hat{\mathbf{a}}$ in $O$.

## 2.7. Spatial Acceleration

The absolute spatial acceleration of a rigid body is just the absolute derivative of its spatial velocity. However, the spatial acceleration of a rigid body differs somewhat from its conventional acceleration.

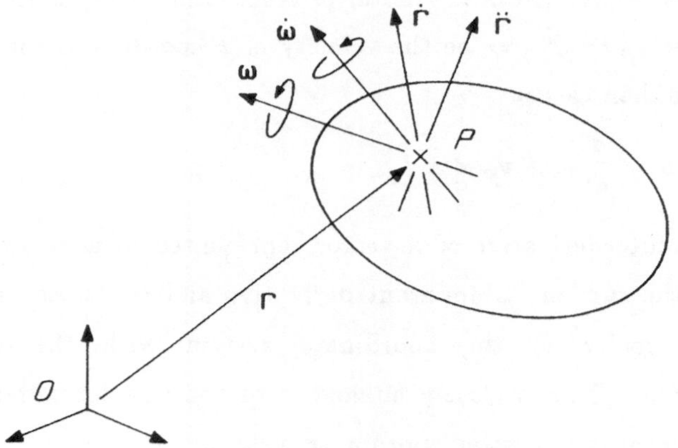


**Figure 2-3:** Acceleration of a rigid body

Consider the rigid body shown in Figure 2-3. It has angular velocity $\omega$, and the linear velocity of a point $P$ in the rigid body is $\dot{\mathbf{r}}$, where $\mathbf{r}$ is the position vector of $P$. The spatial velocity of the body in stationary coordinates at $O$ is

$$\hat{\mathbf{v}}_O = \begin{bmatrix} \omega \\ \dot{\mathbf{r}} + \mathbf{r} \times \omega \end{bmatrix} , \tag{2.28}$$

and its absolute spatial acceleration is the component-wise derivative of this, which is

$$\hat{\mathbf{a}}_O = \begin{bmatrix} \dot{\omega} \\ \ddot{\mathbf{r}} + \dot{\mathbf{r}} \times \omega + \mathbf{r} \times \dot{\omega} \end{bmatrix} . \tag{2.29}$$

The angular component of the spatial acceleration is the angular acceleration of the body, but the linear component is the rate of change over time of the velocity of whichever point in the rigid body happens to be passing through the origin. This differs from the conventional acceleration of a point in a rigid body, which is the acceleration of a point fixed in the body. The conventional acceleration of the body-fixed point at the origin is

$$\begin{aligned} \alpha_O &= \frac{d^2}{dt^2}(\mathbf{r} - \mathbf{r}') \\ &= \frac{d}{dt}(\dot{\mathbf{r}} - \omega \times \mathbf{r}') \\ &= \ddot{\mathbf{r}} - \dot{\omega} \times \mathbf{r} - \omega \times \omega \times \mathbf{r} , \end{aligned}$$

where $\mathbf{r}'$ is a body-fixed vector instantaneously equal to $\mathbf{r}$.

To clarify this concept, consider the rigid body shown in Figure 2.4, which is rotating with constant angular velocity about a fixed axis. The body has constant spatial velocity, so its spatial acceleration must be zero. Every point in the body not on the axis of rotation is following a circular path, and so has non-zero conventional acceleration; but whichever point is passing through the origin has velocity $\mathbf{v}_O$. This velocity is a constant, so the linear part of the spatial acceleration, which is the derivative of this, is zero.

**Example 3**

Continuing with the robot defined in Example 2, let us find the spatial acceleration of the end effector in terms of the scalar joint

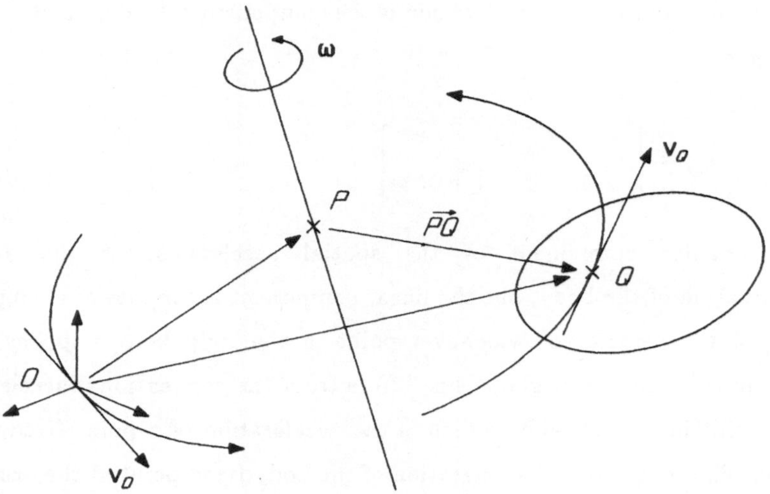


**Figure 2-4:** Acceleration of object with constant angular velocity

accelerations $\ddot{q}_i$. From Example 2, the velocity of the end effector is given by $\hat{\mathbf{v}}_n = \sum_{i=1}^{n} \hat{\mathbf{s}}_i \dot{q}_i$. The acceleration, $\hat{\mathbf{a}}_n$, is the derivative of this, so

$$\hat{\mathbf{a}}_n = \frac{d}{dt} \sum_{i=1}^{n} \hat{\mathbf{s}}_i \dot{q}_i$$

$$= \sum_{i=1}^{n} \left( \frac{d}{dt} \hat{\mathbf{s}}_i \dot{q}_i + \hat{\mathbf{s}}_i \ddot{q}_i \right).$$

Now $\hat{\mathbf{s}}_i$ is moving with velocity $\hat{\mathbf{v}}_i$, so $\frac{d}{dt} \hat{\mathbf{s}}_i = \hat{\mathbf{v}}_i \hat{\times} \hat{\mathbf{s}}_i$ and

$$\hat{\mathbf{a}}_n = \sum_{i=1}^{n} \left( \hat{\mathbf{v}}_i \hat{\times} \hat{\mathbf{s}}_i \dot{q}_i + \hat{\mathbf{s}}_i \ddot{q}_i \right).$$

This is not the most efficient way to calculate the acceleration. A more efficient method is as follows. If $\hat{\mathbf{v}}_i$ is the absolute spatial velocity of link $i$, then

$$\hat{\mathbf{v}}_i = \hat{\mathbf{v}}_{i-1} + \hat{\mathbf{s}}_i \dot{q}_i .$$

Differentiating this equation gives

$$\hat{\mathbf{a}}_i = \hat{\mathbf{a}}_{i-1} + \hat{\mathbf{s}}_i \ddot{q}_i + \hat{\mathbf{v}}_i \hat{\times} \hat{\mathbf{s}}_i \dot{q}_i ,$$

where $\hat{\mathbf{a}}_i$ and $\hat{\mathbf{a}}_{i-1}$ are the absolute spatial accelerations of links $i$ and $i-1$ respectively. Given that $\hat{\mathbf{v}}_0 = \hat{\mathbf{a}}_0 = \hat{\mathbf{0}}$, successive applications of these formulae with $i$ taking values from 1 to $n$ will give the velocity and acceleration of each link in turn.

## 2.8. Screws and Motors

Spatial vectors bear a close resemblance to mathematical quantities called screws and motors, which are used extensively in kinematics. Most of the material in this chapter draws on concepts in the screw and motor algebras. Indeed, the spatial vector algebra can be thought of as an adaptation of screw theory for the purpose of expressing rigid-body dynamics clearly and concisely; although this notion will only be useful to those already familiar with screw theory.

### 2.8.1. Screws

A screw, as defined by Ball [4],[4] is a directed line with which a linear magnitude, called the pitch, is associated. A screw is essentially a geometrical concept. Five numbers are needed to define a screw: four to define the screw axis and one for the pitch. Any finite or infinitesimal displacement of a rigid body can be represented as a twist of a certain magnitude about a screw, which consists of a rotation about the screw axis and a translation along it. The twist magnitude determines the magnitude

[4] This is not the earliest definition of a screw.

of the rotation and the pitch determines the relative magnitude of the translation. A screw of zero pitch gives a pure rotation and one of infinite pitch a pure translation. Similarly, any system of forces acting on a rigid body can be represented as a wrench of a certain magnitude on a screw, which consists of a force acting along the screw axis and a parallel couple. The wrench magnitude gives the magnitude of the force, and the pitch the relative magnitude of the couple. Six numbers are needed to define a twist or a wrench: five for the screw and one for the twist or wrench magnitude.

Screws and finite twists are not vector quantities, but infinitesimal twists, derivatives of twists (twist velocities, etc.) and wrenches are. There is a one-to-one correspondence between twist velocities and spatial velocities, wrenches and spatial forces, and between finite twists and spatial rigid-body transformations.

Screw algebra can be represented in a variety of ways. The geometric approach (which is the one used by Ball) takes the geometric concept of a screw as its primitive entity and represents twists and wrenches as scalars associated with screws. The algebraic approach is to represent twists and wrenches via screw coordinates, which may be a set of six real numbers or a set of three dual numbers (see below). The use of real numbers results in an algebra which is very similar to spatial algebra.

Screws are used extensively in kinematics [28], [7], [77], [81], [82], [17], [36], [42] but less so in dynamics [78] although they were originally devised by Ball to study the dynamics of a rigid body. (Ball's representation of inertia is inferior to the one we will be using.)

### 2.8.2. Motors

The term motor was originally introduced by Clifford [13] to mean a magnitude associated with a screw. He said that a motor consists of a rotor (line vector) part and a vector (free vector) part. Nowadays the term is generally used to mean a vector of three dual numbers -- i.e., a dual vector. A dual number (also introduced by Clifford) is one of the form $a + \epsilon\, b$, where a and b are real numbers and $\epsilon$ is an algebraic unit having the formal property that $\epsilon^2 = 0$.[5] The motor representation of a spatial vector $[\, \mathbf{a}^T \mathbf{a}_O{}^T ]^T$ is $\mathbf{a} + \epsilon\, \mathbf{a}_O$.

The usefulness of dual numbers derives from the principle of transference [17], [49] which allows one to obtain valid motor formulae from the corresponding vector formulae by the process of dualisation -- the replacement of real numbers representing distances, angles, etc. in the formula with dual numbers. Thus the motor algebra may readily be obtained from the vector algebra, and since there is a one-to-one correspondence between motors and screws with magnitudes, the motor algebra offers a very convenient way to approach the screw algebra. This is the approach taken by Dimentberg, who speaks of reducing a screw to a point P as representing it in the form $\mathbf{a} + \epsilon\, \mathbf{a}_P$. However, the principle of transference only applies if the functions and operators are analytic (in the same sense as for functions of a complex variable), and since the only representation of inertia available in dual numbers is the inertia binor (see Chapter 3), which is not analytic, the principle of transference does not apply to dynamic equations.

The motor and dual-number algebras are described by Dimentberg [17]and Brand [11] and to a lesser extent by Bottema and Roth [7] (who call motors dual vectors) and Hunt [28]. Dual numbers are also used extensively

---

[5] The symbol $\omega$ is often used in place of $\epsilon$.

in kinematics [81], [82], [80], [18] but much less so in dynamics [79]. The main obstacle to their use in dynamics is the difficulty of representing inertias.

There is one important difference between the use of dual numbers and real numbers to represent screw algebra: the scalar product of two dual vectors produces a dual number result, whilst the scalar product of two real vectors produces a real number result.

Mises [69], [70] used 6-element real vectors to represent motors, and developed a real-number motor algebra, including motor dyadics, which is very similar to spatial vector algebra. The main difference between the two is that spatial vector algebra uses the operators of matrix algebra plus the spatial transpose operator (defined in Chapter 3), whereas Mises used a number of specially defined operators.

In summary, the geometric approach to screw algebra leads to the representation of vector quantities, such as wrenches and twist velocities, in terms of non-vector quantities, resulting in a notation which is unsuitable for vectorial dynamics; whereas the use of dual numbers makes representing and dealing with inertias difficult. Screws represented in real-number screw coordinates and Mises' real-number motors are both very similar to spatial vectors, and are more suitable for expressing rigid-body dynamics.

# Chapter 3
# Spatial Dynamics

## 3.1. Introduction

The previous chapter introduced the basic notions of spatial quantities and dealt with the kinematic aspects of spatial algebra. This chapter deals with the dynamic aspects -- principally inertia. The spatial momentum of a rigid body is defined, then spatial rigid-body inertia is defined as a mapping between velocity and momentum, and its representation as a $6 \times 6$ matrix is deduced. The basic operations of transformation, differentiation and combination (i.e., addition) of spatial rigid-body inertias are described, and the equations of motion for a rigid body are given. The concepts of inverse inertias and articulated-body inertias are introduced, and their properties and uses are described. The discussion on inertias is concluded with a brief review of alternative representations of inertia. The ease with which inertias can be manipulated is one of the key features of spatial notation.

The remaining sections deal with spatial vector analysis. A scalar product is defined for spatial vectors, which leads to the definition of the spatial transpose operator; then various techniques are described for resolving vectors and projecting them into sub-spaces using the scalar product. The chapter concludes with a summary of spatial vector algebra.

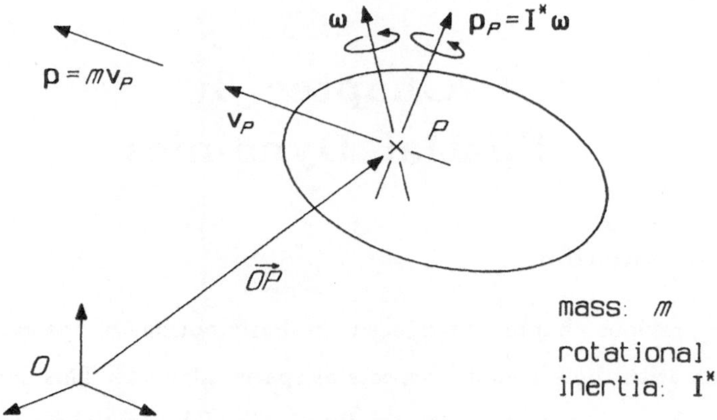


**Figure 3-1:** Spatial momentum of a rigid body

## 3.2. Spatial Rigid-Body Momentum

Consider the rigid body shown in Figure 3-1, which has a mass of $m$, position of centre of mass at $P$, and a rotational inertia of $\mathbf{I}^*$ about the centre of mass. The body has a spatial velocity of $\hat{\mathbf{v}}_O = [\,\omega^T\, \mathbf{v}_O{}^T\,]^T$, and the linear velocity of the centre of mass is $\mathbf{v}_P = \mathbf{v}_O + \vec{PO} \times \omega$. The linear momentum of the rigid body is a line vector with magnitude and direction given by $\mathbf{P} = m\,\mathbf{v}_P$ and line of action passing through the body's centre of mass. It can be expressed as a spatial vector $\hat{\mathbf{P}}_O = [\,\mathbf{P}^T\ (\vec{OP} \times \mathbf{P})^T\,]^T$. The vector $\vec{OP} \times \mathbf{P}$ is the body's moment of momentum about the origin. The body also has an intrinsic angular momentum of $\mathbf{I}^*\,\omega$, which behaves like a free vector and can be expressed as a spatial vector $\hat{\mathbf{P}}^* = [\,\mathbf{0}^T\ (\mathbf{I}^*\,\omega)^T\,]^T$. The total momentum of the body is the sum of the linear and angular momenta, and is given by

$$\hat{\mathbf{P}} = \begin{bmatrix} m\mathbf{v}_P \\ \vec{OP} \times m\mathbf{v}_P \end{bmatrix} + \begin{bmatrix} \mathbf{0} \\ \mathbf{I}^*\omega \end{bmatrix}$$

$$= \begin{bmatrix} m(\mathbf{v}_O + \vec{PO} \times \omega) \\ \mathbf{I}^*\omega + \vec{OP} \times m(\mathbf{v}_O + \vec{PO} \times \omega) \end{bmatrix} \tag{3.1}$$

Yang's approach to dynamic analysis [79] is based on this equation, but we must go one step further and obtain an explicit expression for the inertia.

## 3.3. Spatial Rigid-Body Inertia

Let the spatial inertia of a rigid body be defined as the tensor which transforms the spatial velocity of a rigid body into its spatial momentum. We postulate a linear equation of the form $\hat{\mathbf{P}} = \hat{\mathbf{I}}\,\hat{\mathbf{v}}$, where $\hat{\mathbf{I}}$ is the spatial inertia of the rigid body and is represented by a $6 \times 6$ matrix. From Equation (3.1) we have

$$\hat{\mathbf{P}} = \begin{bmatrix} m\,\vec{PO} \times \omega + m\,\mathbf{v}_O \\ (\,\mathbf{I}^* + \vec{OP} \times m\,\vec{PO} \times\,)\omega + \vec{OP} \times m\,\mathbf{v}_O \end{bmatrix}$$

$$= \begin{bmatrix} m\,\vec{PO} \times & m\,\mathbf{1} \\ \mathbf{I}^* + \vec{OP} \times m\,\vec{PO} \times & \vec{OP} \times m \end{bmatrix} \hat{\mathbf{v}}_O$$

$$= \hat{\mathbf{I}}_O\,\hat{\mathbf{v}}_O\ , \tag{3.2}$$

so the spatial inertia is

$$\hat{\mathbf{I}}_O = \begin{bmatrix} m\,\vec{PO} \times & m\,\mathbf{1} \\ \mathbf{I}^* + \vec{OP} \times m\,\vec{PO} \times & \vec{OP} \times m \end{bmatrix} . \tag{3.3}$$

This matrix can be expressed more concisely with the introduction of the $3 \times 3$ matrices $\mathbf{M} = m\,\mathbf{1}$, $\mathbf{H} = \vec{OP} \times m$ and $\mathbf{I} = \mathbf{I}^* + \vec{OP} \times m\,\vec{PO} \times$, whence

$$\hat{\mathbf{I}}_O = \begin{bmatrix} \mathbf{H}^T & \mathbf{M} \\ \mathbf{I} & \mathbf{H} \end{bmatrix} . \tag{3.4}$$

**M**, **H** and **I** are matrix representations of the zeroth, first and second moments of mass of the rigid body about the origin: the zeroth moment is just the mass; the first moment is $\mathbf{h} = m\,\vec{OP}$, and $\mathbf{H} = \mathbf{h}\times$; and the second moment is the rotational inertia of the rigid body about the origin, which is equal to **I** by the parallel axes theorem ($\vec{OP}\times m\,\vec{PO}\times$ being the rotational inertia about the origin of a point mass $m$ at $P$). **M** and **I** are symmetric.

If the coordinate origin coincides with the rigid body's centre of mass, then $\vec{OP} = \mathbf{0}$ and the spatial inertia takes on the special form

$$\hat{\mathbf{I}}_O = \begin{bmatrix} \mathbf{0} & \mathbf{M} \\ \mathbf{I}^* & \mathbf{0} \end{bmatrix} .$$

### 3.3.1. Properties of Spatial Rigid-Body Inertias

Unlike rotational inertia, the matrix representing spatial inertia is neither symmetric nor positive definite in the conventional sense. However, when the spatial transpose operator is introduced the spatial inertia matrix will be found to be symmetric and positive definite with respect to that operator.

The rigid-body inertia matrix is always non-singular, for if it were singular then it would be possible for the rigid body to have zero momentum with non-zero velocity.

The coordinate transformation rule for spatial inertia is determined from its definition as a linear mapping of spatial vectors. If $\hat{\mathbf{P}}_P$, $\hat{\mathbf{I}}_P$ and $\hat{\mathbf{v}}_P$ are the momentum, inertia and velocity of a rigid body represented in coordinates $P$, and $\hat{\mathbf{P}}_O$, $\hat{\mathbf{I}}_O$ and $\hat{\mathbf{v}}_O$ are the same represented in coordinates $O$, then $\hat{\mathbf{P}}_P = \hat{\mathbf{I}}_P\,\hat{\mathbf{v}}_P$ and $\hat{\mathbf{P}}_O = \hat{\mathbf{I}}_O\,\hat{\mathbf{v}}_O$. Now $\hat{\mathbf{P}}_P = {}_P\hat{\mathbf{X}}_O\,\hat{\mathbf{P}}_O$ and $\hat{\mathbf{v}}_O = {}_O\hat{\mathbf{X}}_P\,\hat{\mathbf{v}}_P$, where ${}_P\hat{\mathbf{X}}_O$ and ${}_O\hat{\mathbf{X}}_P$ are the usual vector transformation matrices, so

$$\hat{\mathbf{I}}_P\,\hat{\mathbf{v}}_P = \hat{\mathbf{P}}_P = {}_P\hat{\mathbf{X}}_O\,\hat{\mathbf{P}}_O = {}_P\hat{\mathbf{X}}_O\,\hat{\mathbf{I}}_O\,\hat{\mathbf{v}}_O$$

$$= {}_P\hat{\mathbf{X}}_O\,\hat{\mathbf{I}}_O\,{}_O\hat{\mathbf{X}}_P\,\hat{\mathbf{v}}_P\,,$$

and

$$\hat{\mathbf{I}}_P = {}_P\hat{\mathbf{X}}_O\,\hat{\mathbf{I}}_O\,{}_O\hat{\mathbf{X}}_P\,. \tag{3.5}$$

This is just the usual tensor transformation rule, and applies to any quantity defined as a linear mapping of vectors.

The derivative with respect to time of a spatial inertia represented in a stationary coordinate frame is just the component-wise derivative. A formula for differentiating a spatial inertia represented in a moving coordinate frame is most easily obtained via the transformation approach described in Chapter 2. If $P$ is a coordinate system moving with velocity $\hat{\mathbf{v}}$, $O$ a stationary one, and $\hat{\mathbf{I}}$ the inertia to be differentiated, then, using subscripts to indicate the coordinate system of representation, we have from Equation (3.5) that

$$\frac{d}{dt}\hat{\mathbf{I}}_P = {}_P\hat{\mathbf{X}}_O\,\frac{d'}{dt}\hat{\mathbf{I}}_O\,{}_O\hat{\mathbf{X}}_P\,.$$

But

$$\frac{d'}{dt}\hat{\mathbf{I}}_O = \frac{d'}{dt}(\,{}_O\hat{\mathbf{X}}_P\,\hat{\mathbf{I}}_P\,{}_P\hat{\mathbf{X}}_O\,)$$

$$= {}_O\hat{\mathbf{X}}_P\,\hat{\mathbf{v}}_P\,\hat{\times}\,\hat{\mathbf{I}}_P\,{}_P\hat{\mathbf{X}}_O + {}_O\hat{\mathbf{X}}_P\,\frac{d'}{dt}\hat{\mathbf{I}}_P\,{}_P\hat{\mathbf{X}}_O - {}_O\hat{\mathbf{X}}_P\,\hat{\mathbf{I}}_P\,\hat{\mathbf{v}}_P\,\hat{\times}\,{}_P\hat{\mathbf{X}}_O$$

(using Equation (2.25) and Equation (2.13) applied to spatial cross), so

$$\frac{d}{dt}\hat{\mathbf{I}}_P = \frac{d'}{dt}\hat{\mathbf{I}}_P + \hat{\mathbf{v}}_P\,\hat{\times}\,\hat{\mathbf{I}}_P - \hat{\mathbf{I}}_P\,\hat{\mathbf{v}}_P\,\hat{\times}\,. \tag{3.6}$$

If a rigid body has inertia $\hat{\mathbf{I}}$ and is moving with an absolute velocity $\hat{\mathbf{v}}$, then the absolute derivative of $\hat{\mathbf{I}}$ is given by

$$\frac{d}{dt}\hat{\mathbf{I}} = \hat{\mathbf{v}}\hat{\times}\hat{\mathbf{I}} - \hat{\mathbf{I}}\,\hat{\mathbf{v}}\hat{\times}\,. \tag{3.7}$$

This can be verified from Equation (3.6).

The inertia of a composite rigid body is the sum of the inertias of its components. If a body $b$ is a rigid assembly composed of rigid bodies $b_i$ then the inertia, $\hat{\mathbf{I}}$, of $b$ is given in terms of the inertias, $\hat{\mathbf{I}}_i$, of $b_i$ by

$$\hat{\mathbf{I}} = \sum_i \hat{\mathbf{I}}_i\,. \tag{3.8}$$

This single spatial equation takes the place of three equations in the conventional approach: one to compute the composite mass, one to compute the composite centre of mass, and one to compute the rotational inertia about the new centre of mass.

## 3.4. The Equations of Motion

In the absence of an applied force, the momentum of a rigid body is conserved; so

$$\hat{\mathbf{I}}\,\hat{\mathbf{v}} = \text{ constant }\,. \tag{3.9}$$

If a force is applied, then the rate of change of momentum is equal to the net applied force:

$$\hat{\mathbf{f}} = \frac{d}{dt}(\hat{\mathbf{I}}\,\hat{\mathbf{v}})$$

$$= \hat{\mathbf{I}}\,\hat{\mathbf{a}} + \hat{\mathbf{v}}\hat{\times}\hat{\mathbf{I}}\,\hat{\mathbf{v}}, \tag{3.10}$$

where $\hat{\mathbf{a}}$ is the rigid body's spatial acceleration. It will often be convenient to regard Equation (3.10) as an inhomogeneous linear equation

$$\hat{\mathbf{f}} = \hat{\mathbf{I}}\,\hat{\mathbf{a}} + \hat{\mathbf{p}}, \tag{3.11}$$

where $\hat{\mathbf{p}}$ is the force which must be applied to the rigid body to produce zero spatial acceleration, and is called the bias force. Equation (3.9) combines the laws of conservation of linear and angular momentum, and Equation (3.10) combines Newton's and Euler's equations for linear and angular motion of a rigid body.

**Example 1**

In Chapter 2, Section 7, we learned that a rigid body traversing a circular course with constant speed has constant spatial velocity and hence zero spatial acceleration; but a force is required to keep the body on this course, and we are now able to calculate it:

$$\hat{\mathbf{f}} = \hat{\mathbf{I}}\,\hat{\mathbf{a}} + \hat{\mathbf{v}} \times \hat{\mathbf{I}}\,\hat{\mathbf{v}}$$

$$= \hat{\mathbf{v}} \times \hat{\mathbf{I}}\,\hat{\mathbf{v}}.$$

$\hat{\mathbf{f}}$ will only be zero if $\hat{\mathbf{v}}$ is zero, if $\hat{\mathbf{v}}$ is purely linear, or if $\hat{\mathbf{I}}\,\hat{\mathbf{v}}$ is coaxial with $\hat{\mathbf{v}}$. This last case will occur if the screw axis of $\hat{\mathbf{v}}$ is one of the principal axes of the rotational inertia of the rigid body.

## 3.5. Inverse Inertia

The inverse inertia of a rigid body relates momentum to velocity. If $\hat{\mathbf{P}} = \hat{\mathbf{I}}\,\hat{\mathbf{v}}$ then $\hat{\mathbf{v}} = \hat{\Phi}\,\hat{\mathbf{P}}$, where $\hat{\Phi} = \hat{\mathbf{I}}^{-1}$ is the inverse inertia of the rigid body. Inverse inertias obey the same rules for transformation and differentiation as do inertias, but their composition to form the inverse inertia of a composite body is accomplished by the harmonic sum, thus:

$$\hat{\Phi} = \Big(\sum_i \hat{\Phi}_i^{-1}\Big)^{-1}. \tag{3.12}$$

If a force $\hat{\mathbf{f}}$ is applied to a constrained body producing an acceleration $\hat{\mathbf{a}}$, then the acceleration may be expressed as

$$\hat{\mathbf{a}} = \hat{\Phi}\,\hat{\mathbf{f}} + \hat{\mathbf{b}}, \tag{3.13}$$

where $\hat{\Phi}$ is the apparent inverse inertia of the body, and $\hat{\mathbf{b}}$ is a vector depending on the velocity, motion constraints, etc., and is called the bias acceleration.

The main advantage in using inverse inertias comes in dealing with bodies constrained to have fewer than six degrees of motion freedom. For such a body, the equation $\hat{\Phi}\,\hat{\mathbf{f}} = \hat{\mathbf{0}}$ has non-trivial solutions, so $\hat{\Phi}$ is singular and $\hat{\mathbf{I}}$ does not exist. In fact the rank of the inverse inertia equals the number of degrees of motion freedom possessed by the body, and the solutions to $\hat{\Phi}\,\hat{\mathbf{f}} = \hat{\mathbf{0}}$ form the space of possible reaction forces generated by the motion constraints. The main disadvantage of inverse inertias is the inconvenience of the combination rule: the harmonic sum is difficult to perform, especially if some of the matrices are singular.

The form of an unconstrained rigid-body inverse inertia with mass $m$, centre of mass at $P$ and central inertia tensor $\mathbf{I}^*$ is

$$\hat{\Phi} = \begin{bmatrix} \mathbf{I}^{*-1}\vec{PO}\times & \mathbf{I}^{*-1} \\ 1/m\ \mathbf{1}+\vec{OP}\times\ \mathbf{I}^{*-1}\vec{PO}\times & \vec{OP}\times\ \mathbf{I}^{*-1} \end{bmatrix} . \tag{3.14}$$

This equation is identical to Equation (3.2) with $\frac{1}{m}\mathbf{1}$ replacing $\mathbf{I}^*$ and $\mathbf{I}^{*-1}$ replacing $m\ \mathbf{1}$.

## 3.6. Articulated-Body Inertias

An articulated body is a collection of rigid bodies connected by joints. All interactions between an articulated body and the rest of a rigid-body system must occur through a single member of the articulated body, which is called the *handle*. An articulated-body inertia gives the relationship between a force applied to the handle (by the rest of the system) and its acceleration, taking into account the dynamics of the articulated body. The relationship is linear and may be expressed in the form

$$\hat{\mathbf{f}} = \hat{\mathbf{I}}^A\,\hat{\mathbf{a}} + \hat{\mathbf{p}}, \tag{3.15}$$

where $\hat{\mathbf{f}}$ is the applied force, also known as the test force, $\hat{\mathbf{a}}$ is the handle's acceleration, $\hat{\mathbf{I}}^A$ is the articulated-body inertia and $\hat{\mathbf{p}}$ is its associated bias force, which is the force which must be applied to the handle to give it zero acceleration.

The basic idea of the articulated-body inertia is that it allows the group of bodies making up the articulated body to be treated as if they were a single rigid-body-like element of the system. This is possible because all interactions with the rest of the system occur through the handle, and the form of Equation (3.15) is the same as the rigid-body equation of motion, Equation (3.11). The use of articulated-body inertias allows the simplification of a rigid-body system by reducing the apparent number of elements. This is the basic idea behind the articulated-body dynamics algorithm described in Chapter 6.

Equation (3.15) assumes that the handle has six degrees of motion freedom. If this is not the case then the more general equation

$$\hat{\mathbf{a}} = \hat{\Phi}^A \hat{\mathbf{f}} + \hat{\mathbf{b}} \tag{3.16}$$

is required, where $\hat{\Phi}^A$ is the articulated-body inverse inertia and $\hat{\mathbf{b}}$ is the acceleration of the handle in the absence of a test force. An equation of this form always exists.

It is possible to generalise the notion of an articulated body to allow more than one handle. In this case the relationship between a test force applied to one handle and the acceleration of another would be described by a cross-inertia, which would normally need to be expressed as an inverse inertia. For more details on this topic see Chapter 9.

Articulated-body inertias depend only on the rigid-body inertias of the members and the instantaneous kinematics of the connections between the members. Velocity effects and the various forces acting on and within the articulated body affect only the bias force.

### 3.6.1. Properties of Articulated-Body Inertias

Unlike rigid-body inertias, which are defined as linear mappings from velocity to momentum, articulated-body inertias are defined in terms of a linear relationship between a force and an acceleration. An expression of the form $\hat{\mathbf{I}}^A \hat{\mathbf{v}}$, where $\hat{\mathbf{I}}^A$ is an articulated-body inertia and $\hat{\mathbf{v}}$ is a velocity, is meaningless. Articulated-body inertias are tensors and obey the same rules for transformation of representation and differentiation in moving coordinates as do rigid-body inertias. They also have the same physical dimensions as rigid-body inertias.

Articulated-body inertias, when they exist, are symmetric,[6] positive-definite matrices (with respect to spatial transpose); articulated-body inverse inertias are symmetric, positive definite or semi-definite; and inverse cross-inertias are generally non-symmetric, indefinite and singular.

Articulated-body inertias are general symmetric matrices, not conforming to the special form of rigid-body inertias. Physically, this implies that there is no such thing as a centre of mass for an articulated body and that the apparent mass has directional properties analogous to those of rotational inertia. To see this consider the simple articulated body shown in Figure 3-2. It consists of two bodies: a sphere of mass $m_1$ with a cylindrical shaft through the middle, and a cylinder of mass $m_2$ which fits in the shaft. The shaft is parallel to the $y$ axis and the two bodies are connected kinematically by a cylindric joint. Both bodies are initially at rest and their centres of mass coincide at the centre of the sphere.

A force $\hat{\mathbf{f}}_z$ applied to the sphere along a line parallel to the $z$ axis and passing through the centre of the sphere causes both bodies to accelerate with acceleration $[\, \mathbf{0}^T \;\; 1/(m_1+m_2)\, \mathbf{f}_z{}^T \,]^T$, but a force $\hat{\mathbf{f}}_y$ applied to the sphere in the $y$ direction through the centre of the sphere causes only the sphere to

[6] This is certainly true for loop-free articulated bodies with no kinematic contact with ground, but I have not been able to prove this statement for the general case.

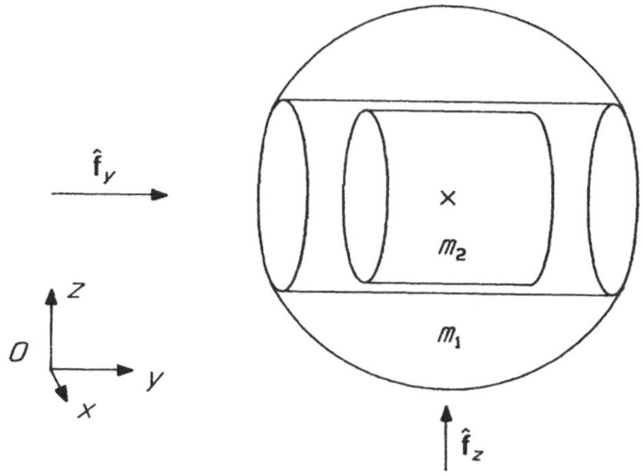


**Figure 3-2:** Articulated body with directional apparent mass

accelerate, so its acceleration is $[\,\mathbf{0}^T \;\; 1/m_1\,\mathbf{f}_y{}^T\,]^T$. In fact the $3 \times 3$ matrix in the mass quadrant of the inertia for this articulated body, taking the sphere as the handle, is

$$\mathbf{M} = \begin{bmatrix} m_1{+}m_2 & 0 & 0 \\ 0 & m_1 & 0 \\ 0 & 0 & m_1{+}m_2 \end{bmatrix} .$$

If the shaft were not parallel to one of the coordinate axes then $\mathbf{M}$ would not be diagonal, but it would still be symmetric. This articulated body exhibits a centre of mass, since any pure force applied to the sphere in a line passing through its centre imparts a purely linear acceleration to it, but if the cylindric joint is replaced by a screw joint then $\hat{\mathbf{f}}_y$ causes the sphere to rotate, and a centre of mass no longer exists.

### 3.6.2. The Calculation of Articulated-Body Inertias

The calculation of articulated-body inertias (and cross-inertias) is very difficult in the general case; indeed only the inverse inertias can be guaranteed to exist. However, if the connectivity of the articulated-body is such that there are no kinematic loops and no connections to ground (an unmovable frame) then the calculation becomes relatively easy. The basic calculation method for this case is described in Chapter 6. The presence of kinematic loops introduces the need to calculate inverse cross-inertias, and the presence of kinematic connections to ground causes some members to have less than six degrees of motion freedom, also necessitating the use of inverse inertias. A calculation method for these cases is described in Chapter 9.

## 3.7. Other Representations for Inertia

Ball [4] analysed the spatial motion of a constrained rigid body using screws. He represented the spatial inertia in terms of a mass and a number of principal screws of inertia. The principal screws are effectively 'eigenscrews' of inertia, and are analogous to the principal axes of rotational inertia. The number of screws required equals the number of degrees of motion freedom possessed by the rigid body. This approach has some advantages in dealing with the dynamics of a single constrained body. Its disadvantages are that it needs a variable number of quantities of two different types (scalar and screw) to describe a spatial inertia, and that combining inertias to form composite inertias is very difficult. Ball extends his analysis to multi-body systems with the theory of screw chains, but the method of analysis is unsuitable for computer implementation.

Spatial rigid-body inertia may be expressed as a $4 \times 4$ symmetric matrix, which is used, in conjunction with $4 \times 4$ transformation matrices, to express the kinetic energy of a moving rigid body. This representation has become popular for use in the Lagrangian approach to robot and mechanism dynamics, where finding an expression for the kinetic energy of the system is one of the first steps; and it was originally introduced by Uicker for this purpose [66], [67], [46], [22], [75]. However, this representation of inertia does not appear to be used for any other purpose, and it does not appear to be applicable to vectorial mechanics. Furthermore, it is not possible to represent articulated-body inertias this way.

Spatial inertia can not be represented directly in terms of dual numbers since the mass part of the inertia transforms the dual part of the velocity motor into (part of) the principal part of the momentum motor. This problem can be surmounted by defining a quantity called a binor [17], which is an ordered pair of $3 \times 3$ matrices of dual numbers. The operation of multiplying the binor $(\mathbf{A}, \mathbf{A}^+)$ by a motor $\mathbf{a} + \epsilon\, \mathbf{a}_O$ is defined as

$$(\mathbf{A}, \mathbf{A}^+)\,(\mathbf{a} + \epsilon\, \mathbf{a}_O) = \mathbf{A}\,\mathbf{a} + \mathbf{A}^+\,\mathbf{a}_O\,.$$

The form of the inertia binor is then $(\mathbf{H}^T + \epsilon\,\mathbf{I}, \mathbf{M} + \epsilon\,\mathbf{H})$. Binors are linear operators, but not analytic. It is therefore not possible to arrive at the inertia binor via the principle of transference. One needs to define the operation of binor transformation, since motor transformation matrices will not work on a binor, and one runs into difficulty when trying to differentiate a binor. Compared with the $6 \times 6$ matrix, inertia binors are very awkward to use.

It is possible to perform dynamic analysis using motors without the need to represent inertia as a quantity in its own right by dealing directly with rigid-body momenta, as done by Yang [79], or by using methods which do not require a vectorial equation of motion, as done by Woo and Freudenstein [78]. Most of the algorithms presented in this book do need to operate explicitly on inertias.

Mises' motor dyadic [69] is a $6 \times 6$ matrix of real numbers that differs from our representation only in the arrangement of the quadrants. Mises' inertia dyadic is given by

$$\begin{bmatrix} \mathbf{M} & \mathbf{H}^T \\ \mathbf{H} & \mathbf{I} \end{bmatrix},$$

which is symmetric in the conventional sense. The difference is accounted for by Mises' definition of special multiplication operators between dyadics and motors. Our arrangement fits neatly into the framework of matrix algebra, requiring only one new operator: the spatial transpose.

## 3.8. The Spatial Scalar Product

So far we have not defined a scalar product on spatial vectors. Such an operation needs to be able to combine the linear and angular components of two spatial vectors in such a way as to produce a single scalar that has some physical significance. The only sensible choice is for the scalar to be the work done by a spatial force over a spatial displacement, or to be more accurate the virtual work done by the force over an infinitesimal displacement, since finite displacements are not vectors. This is the definition used by Mises for his scalar product of motors, and is also the same as the dual part of the scalar product of motors in dual-number algebra [11], [17] and the virtual coefficient of two screws [4].

A linear force $\mathbf{f}$ acting over an infinitesimal linear displacement $\delta$ performs an infinitesimal amount of work (virtual work) $\mathbf{f}\cdot\delta$. Similarly a couple $\tau$ acting over an infinitesimal angular displacement $\theta$ performs an infinitesimal amount of work $\tau\cdot\theta$. Work is a scalar, so it is permissible to add work due to linear and angular effects.

Now let a spatial force $\hat{\mathbf{f}} = [\, \mathbf{f}^T \mathbf{f}_O{}^T \,]^T$ act over an infinitesimal displacement $\hat{\delta} = [\, \delta^T \delta_O{}^T \,]^T$. $\mathbf{f}$ acts through the origin, and the linear displacement of the origin is $\delta_O$, so the linear component of work is $\mathbf{f}\cdot\delta_O$. Similarly, the angular component of work is $\mathbf{f}_O\cdot\delta$, so we define the spatial scalar product as follows:

$$\hat{\mathbf{f}} \hat{\cdot} \hat{\delta} = \mathbf{f} \cdot \delta_O + \mathbf{f}_O \cdot \delta \,. \tag{3.17}$$

This is a true scalar, independent of the choice of origin, as can easily be verified by substituting $\mathbf{f}_P$ and $\delta_P$ in place of $\mathbf{f}_O$ and $\delta_O$ and expanding in terms of $\mathbf{f}_O$, $\delta_O$ and $\vec{OP}$.

This scalar product is defined only between a force-type vector and a motion-type vector.[7] This is the *applicability restriction* on the spatial scalar product. The spatial scalar product of a vector with itself is not defined, so we can not use the normal definition of magnitude for a spatial vector. Two vectors satisfying $\hat{\mathbf{a}} \hat{\cdot} \hat{\mathbf{b}} = 0$ are said to be orthogonal. (In screw theory they are said to be reciprocal.)

An important property of this scalar product is that it is not positive definite,[8] i.e., if we override the applicability restriction and allow the expression $\hat{\mathbf{a}} \hat{\cdot} \hat{\mathbf{a}}$ we find that it may be positive, negative, or zero for non-zero $\hat{\mathbf{a}}$, so we still can not use the normal definition of magnitude. The effects of this property are largely overshadowed by the effects of the applicability restriction.

---

[7] A force-type vector is one where the linear component is the line vector and a motion-type vector is one where the angular component is the line vector. There is a distinction between the two similar to that between polar and axial vectors, for if we change the rotation convention to the left-handed corkscrew then the representation of the spatial vector $\hat{\mathbf{a}}$ becomes either $\hat{\mathbf{a}}'$ or $-\hat{\mathbf{a}}'$ depending on whether it represents a force-type vector or a motion-type vector.

[8] See Lang [35] (Chapter 6) for the definition and properties of such a scalar product.

### 3.9. The Spatial Transpose Operator

The scalar product operation can be expressed in matrix form as

$$\hat{\mathbf{a}} \hat{\cdot} \hat{\mathbf{b}} = \hat{\mathbf{a}}^T \begin{bmatrix} 0 & 1 \\ 1 & 0 \end{bmatrix} \hat{\mathbf{b}},$$

but we will eliminate the intermediate matrix by introducing the spatial transpose operator, $^S$, defined on a spatial vector as follows:

$$\hat{\mathbf{a}}^S = \begin{bmatrix} \mathbf{a} \\ \mathbf{a}_O \end{bmatrix}^S = [\mathbf{a}_O^T \; \mathbf{a}^T]. \tag{3.18}$$

The scalar product is then represented by

$$\hat{\mathbf{a}} \hat{\cdot} \hat{\mathbf{b}} = \hat{\mathbf{a}}^S \hat{\mathbf{b}}. \tag{3.19}$$

The spatial transpose operator is to have the same algebraic properties as the ordinary transpose operator, so $\hat{\mathbf{a}}^{SS} = \hat{\mathbf{a}}$, $(\hat{\mathbf{a}} + \hat{\mathbf{b}} + \dots)^S = \hat{\mathbf{a}}^S + \hat{\mathbf{b}}^S + \dots$, and so on. The spatial transpose of a scalar is just the original scalar, and the definition of the spatial transpose of a tensor $\hat{\mathbf{A}}$ is obtained by requiring that

$$\hat{\mathbf{a}}^S \hat{\mathbf{A}} \hat{\mathbf{b}} = (\hat{\mathbf{a}}^S \hat{\mathbf{A}} \hat{\mathbf{b}})^S = \hat{\mathbf{b}}^S \hat{\mathbf{A}}^S \hat{\mathbf{a}}$$

for all $\hat{\mathbf{a}}$ and $\hat{\mathbf{b}}$ (since $\hat{\mathbf{a}}^S \hat{\mathbf{A}} \hat{\mathbf{b}}$ is a scalar). This results in the following definition:

$$\begin{bmatrix} \mathbf{A} & \mathbf{B} \\ \mathbf{C} & \mathbf{D} \end{bmatrix}^S = \begin{bmatrix} \mathbf{D}^T & \mathbf{B}^T \\ \mathbf{C}^T & \mathbf{A}^T \end{bmatrix}. \tag{3.20}$$

This can be memorised by observing that it is equivalent to an ordinary transpose about the broken diagonal through **B** and **C** as shown below.

$$\begin{bmatrix} \mathbf{A} & \mathbf{B} \\ \mathbf{C} & \mathbf{D} \end{bmatrix} \begin{matrix} \mathbf{A} \\ \mathbf{C} \end{matrix} \rightarrow \begin{bmatrix} \mathbf{D}^T & \mathbf{B}^T \\ \mathbf{C}^T & \mathbf{A}^T \end{bmatrix} \begin{matrix} \mathbf{D}^T \\ \mathbf{C}^T \end{matrix}.$$

As a welcome bonus, use of the spatial transpose operator clears up some of the anomalies between spatial algebra and ordinary vector algebra. In particular, coordinate transformation matrices are spatially orthogonal, i.e., $\hat{\mathbf{X}}^{-1} = \hat{\mathbf{X}}^S$ (or ${}_P\hat{\mathbf{X}}_O{}^S = {}_O\hat{\mathbf{X}}_P$); the spatial cross product matrix is anti-symmetric, i.e., $(\hat{\mathbf{v}}\hat{\times})^S = -\hat{\mathbf{v}}\hat{\times}$; and spatial rigid-body inertias are both symmetric and positive definite, i.e., $\hat{\mathbf{I}} = \hat{\mathbf{I}}^S$ and $\hat{\mathbf{x}}^S\hat{\mathbf{I}}\hat{\mathbf{x}} > 0$ for all $\hat{\mathbf{x}} \neq \hat{\mathbf{0}}$.

**Example 2**

The kinetic energy of a rigid body with inertia $\hat{\mathbf{I}}$ and velocity $\hat{\mathbf{u}}$ is equal to the amount of work done in accelerating it from rest to $\hat{\mathbf{u}}$, so

$$KE = \int_{\hat{\mathbf{v}}=\hat{\mathbf{0}}}^{\hat{\mathbf{v}}=\hat{\mathbf{u}}} \hat{\mathbf{f}}^S\hat{\mathbf{v}}\,dt\,.$$

Now

$$\hat{\mathbf{f}} = \frac{d}{dt}(\hat{\mathbf{I}}\hat{\mathbf{v}}) = \hat{\mathbf{I}}\hat{\mathbf{a}} + \hat{\mathbf{v}}\hat{\times}\hat{\mathbf{I}}\hat{\mathbf{v}}$$

and

$$\begin{aligned}
\hat{\mathbf{f}}^S\hat{\mathbf{v}} &= 1/2\,(\hat{\mathbf{f}}^S\hat{\mathbf{v}} + \hat{\mathbf{v}}^S\hat{\mathbf{f}}) \\
&= 1/2\,(\frac{d}{dt}(\hat{\mathbf{I}}\hat{\mathbf{v}})^S\hat{\mathbf{v}} + \hat{\mathbf{v}}^S(\hat{\mathbf{I}}\hat{\mathbf{a}} + \hat{\mathbf{v}}\hat{\times}\hat{\mathbf{I}}\hat{\mathbf{v}})) \\
&= 1/2\,(\frac{d}{dt}(\hat{\mathbf{I}}\hat{\mathbf{v}})^S\hat{\mathbf{v}} + \hat{\mathbf{v}}^S\hat{\mathbf{I}}\frac{d}{dt}\hat{\mathbf{v}}) \\
&= 1/2\,\frac{d}{dt}(\hat{\mathbf{v}}^S\hat{\mathbf{I}}\hat{\mathbf{v}}),
\end{aligned}$$

so

$$KE = 1/2\,\hat{\mathbf{u}}^S\hat{\mathbf{I}}\hat{\mathbf{u}}\,.$$

## 3.10. Spatial Vector Analysis

This section covers a number of topics with the general theme of resolving vectors into components on generalised bases. The main topics are the definition of spatial vector sub-spaces, a technique for resolving a vector into components which are elements of given sub-spaces, and the tensor product of spatial vectors.

### 3.10.1. Coordinates, Bases and Sub-Spaces

A set of spatial vectors $\hat{\mathbf{e}}_i$ is said to be linearly independent if the only solution to $\sum a_i \hat{\mathbf{e}}_i = \hat{\mathbf{0}}$ is for all $a_i$ to be zero. Since spatial vector space is 6-dimensional, there can be at most six vectors in a linearly independent set. A set of $r$ linearly independent spatial vectors span an $r$-dimensional sub-space of spatial vector space and define a basis on that sub-space. Any element of the sub-space may be expressed in the form

$$\hat{\mathbf{a}} = \sum_{i=1}^{r} a_i \hat{\mathbf{e}}_i , \qquad (3.21)$$

where the numbers $a_i$ are unique for each $\hat{\mathbf{a}}$ and are its coordinates in the basis $\{\hat{\mathbf{e}}_i\}$. The members of a basis, and hence all members of a sub-space, must be of the same type: motion or force. A standard basis, as defined in Chapter 2, is special in that its vectors are given a physical interpretation as line vectors and free vectors. No physical interpretation exists for the vectors of a general basis except through their representations in a standard basis. Unless a sub-space is indicated, a basis is assumed to span the entire spatial vector space.

Equation (3.21) may be expressed in matrix form as

$$\hat{\mathbf{a}} = \hat{\mathbf{E}}\, \mathbf{a} , \qquad (3.22)$$

where $\hat{\mathbf{E}}$ is a $6 \times r$ matrix whose $i^{th}$ column is $\hat{\mathbf{e}}_i$, and $\mathbf{a}$ is an $r \times 1$ vector whose $i^{th}$ element is $a_i$. $\hat{\mathbf{E}}$ represents both a vector sub-space (the column space of the matrix) and the particular basis $\{\hat{\mathbf{e}}_i\}$ spanning that sub-space. **a** is just a vector of coefficients identifying a particular member of the sub-space.

The matrix $\hat{\mathbf{A}}\hat{\mathbf{E}}$, where $\hat{\mathbf{A}}$ is a general spatial transformation, represents the $r$-dimensional sub-space spanned by $\{\hat{\mathbf{A}}\hat{\mathbf{e}}_i\}$. Any matrix $\hat{\mathbf{E}}\ \mathbf{A}$, where **A** is a non-singular $r \times r$ matrix, is equivalent to $\hat{\mathbf{E}}$ in that it spans the same vector sub-space, but with a different set of basis vectors. The spatial transpose of a sub-space matrix is a straightforward extension of the spatial transpose of a vector, and is given by

$$\hat{\mathbf{E}}^S = \begin{bmatrix} \mathbf{E} \\ \mathbf{E}_O \end{bmatrix}^S = [\mathbf{E}_O^T\ \mathbf{E}^T]\,, \tag{3.23}$$

where $\mathbf{E}$ and $\mathbf{E}_O$ are $3 \times r$ matrices. It is convenient to extend the definition of spatial transpose to apply to any non-spatial vector or matrix. For such a quantity it is defined to be the same operation as ordinary transpose. So, for example,

$$\hat{\mathbf{a}}^S = (\,\hat{\mathbf{E}}\ \mathbf{a}\,)^S = \mathbf{a}^T \hat{\mathbf{E}}^S\,. \tag{3.24}$$

If $\hat{\mathbf{S}}$ is an $r$-dimensional sub-space of the spatial vector space, then $\hat{\mathbf{S}}^{\perp}$ is the sub-space orthogonal to $\hat{\mathbf{S}}$, defined as the space of all vectors orthogonal to every element of $\hat{\mathbf{S}}$. $\hat{\mathbf{S}}^{\perp}$ has dimension $6-r$ and is of the opposite type to $\hat{\mathbf{S}}$. If $\hat{\mathbf{S}}\,\alpha$ is a vector in $\hat{\mathbf{S}}$ and $\hat{\mathbf{S}}^{\perp}\,\beta$ a vector in $\hat{\mathbf{S}}^{\perp}$ then $(\,\hat{\mathbf{S}}\,\alpha\,)^S\,\hat{\mathbf{S}}^{\perp}\,\beta = 0$; but this is true for all $\alpha$ and $\beta$, so $\hat{\mathbf{S}}$ and $\hat{\mathbf{S}}^{\perp}$ must satisfy

$$\hat{\mathbf{S}}^S\,\hat{\mathbf{S}}^{\perp} = \mathbf{0}\,, \tag{3.25}$$

where here $\mathbf{0}$ is an $r \times (6-r)$ matrix of zeroes. Any two sub-spaces which satisfy this equation are said to be orthogonal, regardless of whether their dimensions add up to 6.

Vector sub-space matrices are used in Chapter 7 to represent multiple-degree-of-freedom joints.

### 3.10.2. Resolving Vectors into Sub-Spaces and Components

It often happens that we wish to resolve a vector into its components in two sub-spaces. This is only possible in general if the two sub-spaces span the spatial vector space, and the solution is unique only if the two sub-spaces have no non-zero element in common. In our case it is also necessary for both sub-spaces to be of the same type as the vector to be resolved. So, given an $r$-dimensional sub-space $\hat{\mathbf{S}}$, if we wish to resolve a vector of the same type as the elements of $\hat{\mathbf{S}}$ into unique components in $\hat{\mathbf{S}}$ and $\hat{\mathbf{T}}$, it is necessary and sufficient that $\hat{\mathbf{T}}$ be a $6-r$ dimensional sub-space of the same type as and having no non-zero element in common with $\hat{\mathbf{S}}$.

Let us consider first the special case of $r=1$. Take two vectors $\hat{\mathbf{s}}$ and $\hat{\mathbf{t}}$ such that $\hat{\mathbf{s}}^S\hat{\mathbf{t}} \neq 0$. We wish to express a third vector $\hat{\mathbf{a}}$ as the sum of two vectors $\hat{\mathbf{a}}_1$ and $\hat{\mathbf{a}}_2$, where $\hat{\mathbf{a}}_1 = \hat{\mathbf{s}}\,\alpha$ and $\hat{\mathbf{t}}^S\hat{\mathbf{a}}_2 = 0$. $\hat{\mathbf{a}}_1$ is a member of the one-dimensional sub-space generated by $\hat{\mathbf{s}}$, and $\hat{\mathbf{a}}_2$ is a member of the orthogonal sub-space to $\hat{\mathbf{t}}$. These spaces satisfy the above conditions, since $\hat{\mathbf{t}}^{\perp}$ has dimension 5 and $\hat{\mathbf{s}}$ is not a member of $\hat{\mathbf{t}}^{\perp}$, so we are justified in saying that

$$\hat{\mathbf{a}} = \hat{\mathbf{a}}_1 + \hat{\mathbf{a}}_2 = \hat{\mathbf{s}}\,\alpha + \hat{\mathbf{a}}_2 .$$

The problem is solved once we know $\alpha$. To find $\alpha$ simply premultiply by $\hat{\mathbf{t}}^S$:

$$\hat{\mathbf{t}}^S\hat{\mathbf{a}} = \hat{\mathbf{t}}^S\hat{\mathbf{s}}\,\alpha + \hat{\mathbf{t}}^S\hat{\mathbf{a}}_2 = \hat{\mathbf{t}}^S\hat{\mathbf{s}}\,\alpha ,$$

so

$$\alpha = \frac{\hat{\mathbf{t}}^S\hat{\mathbf{a}}}{\hat{\mathbf{t}}^S\hat{\mathbf{s}}} . \tag{3.26}$$

The denominator is non-zero by hypothesis.

This procedure can be generalised as follows. Let $\hat{\mathbf{S}}$ and $\hat{\mathbf{T}}$ be two $r$-dimensional sub-spaces such that for every member $\hat{\mathbf{S}}\alpha$ of $\hat{\mathbf{S}}$ there exists at least one member $\hat{\mathbf{T}}\beta$ of $\hat{\mathbf{T}}$ satisfying $(\hat{\mathbf{T}}\beta)^S\hat{\mathbf{S}}\alpha \neq 0$. In terms of matrices, this means that the $r \times r$ matrix $\hat{\mathbf{T}}^S \hat{\mathbf{S}}$ is non-singular. Now the sub-spaces $\hat{\mathbf{S}}$ and $\hat{\mathbf{T}}^{\perp}$ span the spatial vector space, since the sum of their dimensions is 6 and they have no non-zero element in common, so it is possible to express a vector $\hat{\mathbf{a}}$ as $\hat{\mathbf{a}} = \hat{\mathbf{S}}\, \alpha + \hat{\mathbf{T}}^{\perp}\, \beta$. As before, $\alpha$ may be found by premultiplying by $\hat{\mathbf{T}}^S$ thus:

$$\hat{\mathbf{T}}^S \hat{\mathbf{a}} \;=\; \hat{\mathbf{T}}^S \hat{\mathbf{S}}\, \alpha + \hat{\mathbf{T}}^S \hat{\mathbf{T}}^{\perp} \beta \;=\; \hat{\mathbf{T}}^S \hat{\mathbf{S}}\, \alpha\,,$$

giving

$$\alpha = (\, \hat{\mathbf{T}}^S \; \hat{\mathbf{S}})^{-1}\, \hat{\mathbf{T}}^S \hat{\mathbf{a}}\,. \tag{3.27}$$

This process is very useful for resolving forces acting on a rigid body.

**Example 3**

A rigid body with inertia $\hat{\mathbf{I}}$ is initially at rest and its motion is constrained to the sub-space spanned by $\hat{\mathbf{S}}$. Find the acceleration of the body in response to an applied force $\hat{\mathbf{f}}$. To solve this problem, we hypothesise that $\hat{\mathbf{f}}$ may be split into two components, $\hat{\mathbf{f}}_1$ and $\hat{\mathbf{f}}_2$, where $\hat{\mathbf{f}}_1$ is responsible for the motion and $\hat{\mathbf{f}}_2$ is absorbed by the reaction forces produced by the motion constraints. The acceleration is constrained to lie in $\hat{\mathbf{S}}$, and may therefore be expressed as $\hat{\mathbf{S}}\, \alpha$. If $\hat{\mathbf{f}}_1$ is responsible for causing the motion then $\hat{\mathbf{f}}_1 = \hat{\mathbf{I}}\, \hat{\mathbf{S}}\, \alpha$, and is therefore a member of the sub-space $\hat{\mathbf{I}}\, \hat{\mathbf{S}}$. $\hat{\mathbf{f}}_2$ does no work, so $\hat{\mathbf{S}}^S \hat{\mathbf{f}}_2 = \mathbf{0}$ meaning that $\hat{\mathbf{f}}_2$ is a member of $\hat{\mathbf{S}}^{\perp}$. Now $\hat{\mathbf{S}}^S \hat{\mathbf{I}}\, \hat{\mathbf{S}}$ is non-singular, so the hypothesised decomposition is permissible, and the acceleration is given by

$$\hat{\mathbf{S}}\, \alpha = \hat{\mathbf{S}}\, (\, \hat{\mathbf{S}}^S \hat{\mathbf{I}}\, \hat{\mathbf{S}}\, )^{-1}\, \hat{\mathbf{S}}^S \hat{\mathbf{f}}\,.$$

Notice that $\hat{\mathbf{S}}\, (\, \hat{\mathbf{S}}^S \hat{\mathbf{I}}\, \hat{\mathbf{S}}\, )^{-1}\, \hat{\mathbf{S}}^S$ is the apparent inverse inertia of the constrained body.

### 3.10.3. Reciprocal Bases and Orthogonal Bases

The applicability restriction prevents us from considering bases which are orthogonal with respect to the spatial scalar product, but we can consider reciprocal bases. Given the basis $\{\hat{\mathbf{e}}_i\}$, the vectors of the reciprocal basis $\{\hat{\mathbf{f}}_i\}$ are defined by

$$\hat{\mathbf{e}}_i^S \hat{\mathbf{f}}_j = \begin{cases} 0 & \text{if } i \neq j \\ 1 & \text{if } i = j \end{cases} . \tag{3.28}$$

It is always possible to find a reciprocal basis using a version of the Gram-Schmidt orthogonalisation process which takes into account the applicability restriction. The reciprocal basis is useful for finding the components of a vector on a given basis:

$$\text{if} \quad \hat{\mathbf{v}} = \sum \hat{\mathbf{e}}_i v_i \quad \text{then} \quad v_i = \hat{\mathbf{f}}_i^S \hat{\mathbf{v}} . \tag{3.29}$$

It is possible for a basis to be orthogonal with respect to a generalised inner product defined between vectors of the same type. For example, we can say that the motion-type basis $\{\hat{\mathbf{e}}_i\}$ is orthogonal with respect to the inertia $\hat{\mathbf{I}}$ if $\hat{\mathbf{e}}_i^S \hat{\mathbf{I}} \hat{\mathbf{e}}_j = 0$ for all $i \neq j$, and similarly for a force-type basis and an inverse inertia. Such a basis can always be found using a variant of the Gram-Schmidt orthogonalisation process described in [35]. If the matrix is positive definite then so is the inner product.

### 3.10.4. Inertia Dyads and the Tensor Product

The tensor product, $\hat{\mathbf{a}}\,\hat{\mathbf{b}}$, of two spatial vectors $\hat{\mathbf{a}}$ and $\hat{\mathbf{b}}$ is defined as follows: for any third vector $\hat{\mathbf{c}}$,

$$(\hat{\mathbf{a}}\,\hat{\mathbf{b}}) \,\hat{\cdot}\, \hat{\mathbf{c}} = \hat{\mathbf{a}}\,(\hat{\mathbf{b}} \,\hat{\cdot}\, \hat{\mathbf{c}}) . \tag{3.30}$$

The tensor product of two vectors is represented as the $6 \times 6$ matrix $\hat{\mathbf{a}}\,\hat{\mathbf{b}}^S$. Such a quantity is known as a dyad and is a matrix of rank 1 (rank used in the matrix sense). The sum of $r$ independent dyads is a matrix of rank $r$; a set of dyads $\hat{\mathbf{a}}_i \hat{\mathbf{b}}_i^S$ being independent if the vectors $\hat{\mathbf{a}}_i$ are independent and the vectors $\hat{\mathbf{b}}_i$ are independent. Any linear transformation of spatial vectors can be expressed as the sum of six (or fewer) dyads, for a linear transformation is completely specified by its effect on six linearly independent vectors. For example, if $\{\hat{\mathbf{e}}'_i\}$ is the image of $\{\hat{\mathbf{e}}_i\}$ under some transformation $\hat{\mathbf{A}}$, then the transformation can be expressed as

$$\hat{\mathbf{A}} = \sum_{i=1}^{6} \hat{\mathbf{e}}'_i \hat{\mathbf{f}}_i^S ,$$

where $\{\hat{\mathbf{f}}_i\}$ is the reciprocal basis to $\{\hat{\mathbf{e}}_i\}$.

A rigid-body inertia, or indeed any symmetric articulated-body inertia or inverse inertia, can be represented as the sum of six symmetric inertia dyads. Let $\{\hat{\mathbf{s}}_i\}$ be an orthogonal basis with respect to the inertia $\hat{\mathbf{I}}$, i.e., $\hat{\mathbf{s}}_i^S \hat{\mathbf{I}} \hat{\mathbf{s}}_j = 0$ for all $i \neq j$, then $\hat{\mathbf{I}}$ transforms $\{\hat{\mathbf{s}}_i\}$ to $\{\hat{\mathbf{I}} \hat{\mathbf{s}}_i\}$, and $\{\hat{\mathbf{I}} \hat{\mathbf{s}}_i / \hat{\mathbf{s}}_i^S \hat{\mathbf{I}} \hat{\mathbf{s}}_i\}$ is a reciprocal basis to $\{\hat{\mathbf{s}}_i\}$; so $\hat{\mathbf{I}}$ may be represented as

$$\hat{\mathbf{I}} = \sum_{i=1}^{6} \frac{\hat{\mathbf{I}} \hat{\mathbf{s}}_i \hat{\mathbf{s}}_i^S \hat{\mathbf{I}}}{\hat{\mathbf{s}}_i^S \hat{\mathbf{I}} \hat{\mathbf{s}}_i} . \tag{3.31}$$

Similarly, the inverse inertia may be expressed as

$$\hat{\mathbf{I}}^{-1} = \sum_{i=1}^{6} \frac{\hat{\mathbf{s}}_i \hat{\mathbf{s}}_i^S}{\hat{\mathbf{s}}_i^S \hat{\mathbf{I}} \hat{\mathbf{s}}_i} \tag{3.32}$$

by virtue of the identity $\hat{\mathbf{I}}^{-1} = \hat{\mathbf{I}}^{-1} \hat{\mathbf{I}} \hat{\mathbf{I}}^{-1}$; and the identities $\hat{\mathbf{I}} \hat{\mathbf{I}}^{-1} = \hat{\mathbf{I}}^{-1} \hat{\mathbf{I}} = \hat{\mathbf{1}}$ give rise to

$$\sum_{i=1}^{6} \frac{\hat{\mathbf{I}} \hat{\mathbf{s}}_i \hat{\mathbf{s}}_i^S}{\hat{\mathbf{s}}_i^S \hat{\mathbf{I}} \hat{\mathbf{s}}_i} = \sum_{i=1}^{6} \frac{\hat{\mathbf{s}}_i \hat{\mathbf{s}}_i^S \hat{\mathbf{I}}}{\hat{\mathbf{s}}_i^S \hat{\mathbf{I}} \hat{\mathbf{s}}_i} = \hat{\mathbf{1}} . \tag{3.33}$$

If we make the vectors $\hat{\mathbf{s}}_i$ orthonormal with respect to $\hat{\mathbf{I}}$ then the denominators are all unity and can be removed from the equations.

**Example 4**

From example 3 we learned that the inverse inertia of a body with rigid-body inertia $\hat{\mathbf{I}}$ and constrained to move in an $r$-dimensional sub-space $\hat{\mathbf{S}}$is given by

$$\hat{\Phi} = \hat{\mathbf{S}}\,(\,\hat{\mathbf{S}}^S \hat{\mathbf{I}}\,\hat{\mathbf{S}}\,)^{-1} \hat{\mathbf{S}}^S\,.$$

Since here $\hat{\mathbf{S}}$ represents only the vector sub-space, we may choose the basis as we please, and in particular we may choose the vectors $\hat{\mathbf{s}}_i$ making up $\hat{\mathbf{S}}$ to be orthogonal with respect to $\hat{\mathbf{I}}$. If we do this then $\hat{\mathbf{S}}^S \hat{\mathbf{I}}\,\hat{\mathbf{S}}$ becomes diagonal, so $(\,\hat{\mathbf{S}}^S \hat{\mathbf{I}}\,\hat{\mathbf{S}})^{-1}$ is also diagonal, and we may expand the above equation to give

$$\hat{\Phi} = \sum_{i=1}^{r} \frac{\hat{\mathbf{s}}_i \hat{\mathbf{s}}_i^S}{\hat{\mathbf{s}}_i^S \hat{\mathbf{I}}\,\hat{\mathbf{s}}_i}\,.$$

This shows clearly that $\hat{\Phi}$ is the sum of $r$ dyads, one for each degree of freedom, and hence that the rank of an inverse inertia is the degree of motion freedom of the body to which it refers.

## 3.11. Summary of Spatial Algebra

A spatial vector is an element of a 6-dimensional vector space over the real numbers. It may be interpreted geometrically as having a line vector component and a free vector component. Analogous linear and angular physical quantities (e.g., linear and angular velocity, force and couple, etc.) may be amalgamated to form spatial vectors if it makes physical sense to regard one of them as a line vector and the other as a free vector. A spatial vector representing a physical quantity may be classified as being of either motion type or force type depending on whether the line vector part is the angular or linear component respectively.

A spatial vector is represented as a $6 \times 1$ column vector of components in a standard basis, which comprises three unit line vectors and three unit free vectors. The line vectors form the axes of a Cartesian coordinate frame and the free vectors are parallel to these axes. The representation of a spatial vector whose line-vector part is $\mathbf{a}$ (acting in a line through the origin) and whose free-vector part is $\mathbf{a}_O$ is given by

$$\hat{\mathbf{a}} = \begin{bmatrix} \mathbf{a} \\ \mathbf{a}_O \end{bmatrix} .$$

Spatial quantities are denoted by carets. Normally no notational distinction is made between an object and its representation, but if a distinction is needed for clarity then representations are denoted by tildes.

The basic operations of addition and multiplication by a scalar may be performed on spatial vectors provided there is a physical interpretation for such an operation.

The representation of a spatial vector may be transformed from one standard basis to another by a rigid-body coordinate transformation, which is represented as a $6 \times 6$ matrix. The coordinate transformation corresponding to a translation of the origin by $\mathbf{r}$ is

$$\hat{\mathbf{X}} = \begin{bmatrix} 1 & 0 \\ \mathbf{r}\times^T & 1 \end{bmatrix} ,$$

and that corresponding to the coordinate rotation given by the $3 \times 3$ orthogonal rotation matrix $\mathbf{E}$ is

$$\hat{\mathbf{X}} = \begin{bmatrix} \mathbf{E} & \mathbf{0} \\ \mathbf{0} & \mathbf{E} \end{bmatrix} .$$

Transformations are combined and applied to vectors by matrix multiplication, and are given leading and following subscripts to indicate the representation after and before the transformation respectively. If $O$ and $P$ denote two coordinate systems then

$$\hat{\mathbf{a}}_O = {}_O\hat{\mathrm{X}}_P\, \hat{\mathbf{a}}_P,$$

where $\hat{\mathbf{a}}_O$ and $\hat{\mathbf{a}}_P$ are representations of the vector $\hat{\mathbf{a}}$ in $O$ and $P$ coordinates respectively, and ${}_O\hat{\mathrm{X}}_P$ effects the transformation from $P$ coordinates to $O$ coordinates.

The derivative of a spatial vector represented in a stationary coordinate system is its component-wise derivative. If the vector is represented in a coordinate system moving with velocity $\hat{\mathbf{v}}$ then the derivative is given by

$$\frac{d}{dt}\hat{\mathbf{a}} = \frac{d'}{dt}\hat{\mathbf{a}} + \hat{\mathbf{v}} \hat{\times} \hat{\mathbf{a}},$$

where $\frac{d'}{dt}$ denotes component-wise differentiation.

The scalar product between a motion-type vector and a force-type vector is defined as

$$\begin{bmatrix} \mathbf{f} \\ \mathbf{f}_O \end{bmatrix} \hat{\cdot} \begin{bmatrix} \mathbf{v} \\ \mathbf{v}_O \end{bmatrix} = \mathbf{f} \cdot \mathbf{v}_O + \mathbf{f}_O \cdot \mathbf{v} \;.$$

A scalar product is not defined between two motion-type vectors or between two force-type vectors, and in particular the scalar product of a vector with itself is not defined.

To implement the scalar product as a matrix operation, a new operator, the spatial transpose, is introduced and used in place of ordinary transpose. The spatial transpose of a spatial vector $\hat{\mathbf{a}}$ is denoted by $\hat{\mathbf{a}}^S$, and is defined by

$$\hat{\mathbf{a}}^S = \begin{bmatrix} \mathbf{a} \\ \mathbf{a}_O \end{bmatrix}^S = [\mathbf{a}_O^T \; \mathbf{a}^T],$$

The scalar product is given in terms of the spatial transpose by

$$\hat{\mathbf{f}} \hat{\cdot} \hat{\mathbf{v}} = \hat{\mathbf{f}}^S \hat{\mathbf{v}}.$$

The spatial transpose of a dyadic (e.g., an inertia or a transformation) is

$$\begin{bmatrix} \mathbf{A} & \mathbf{B} \\ \mathbf{C} & \mathbf{D} \end{bmatrix}^S = \begin{bmatrix} \mathbf{D}^T & \mathbf{B}^T \\ \mathbf{C}^T & \mathbf{A}^T \end{bmatrix} .$$

All spatial quantities are subject to spatial transpose, but apart from this the normal rules of matrix algebra apply. The use of spatial transpose makes spatial coordinate transformations orthogonal, spatial rigid-body inertias symmetric, and spatial cross-product matrices anti-symmetric.

The spatial inertia of a rigid body is defined as the tensor which maps the velocity of a rigid body to its momentum. It is represented as a $6 \times 6$ matrix which is spatially symmetric and positive definite. The spatial inertia of a rigid body with mass $m$, centre of mass at $\mathbf{c}$, and rotational inertia $\mathbf{I}^*$ about its centre of mass is given by

$$\hat{\mathbf{I}} = \begin{bmatrix} m\,\mathbf{c}\times^T & m\,\mathbf{1} \\ \mathbf{I}^* + \mathbf{c}\times m\,\mathbf{c}\times^T & \mathbf{c}\times m \end{bmatrix} = \begin{bmatrix} \mathbf{H}^T & \mathbf{M} \\ \mathbf{I} & \mathbf{H} \end{bmatrix} .$$

The quadrants **M**, **H** and **I** contain the zeroth, first and second moments of mass about the origin. Rigid-body inertias may be summed to form the inertia of a composite rigid body. The equation of motion for a rigid body is

$$\hat{\mathbf{f}} = \frac{d}{dt}(\hat{\mathbf{I}}\,\hat{\mathbf{v}}) = \hat{\mathbf{I}}\,\hat{\mathbf{a}} + \hat{\mathbf{v}}\times\hat{\mathbf{I}}\,\hat{\mathbf{v}},$$

where $\hat{\mathbf{f}}$ is the net force acting on the body, $\hat{\mathbf{I}}$ is its inertia, $\hat{\mathbf{v}}$ its velocity, $\hat{\mathbf{I}}\,\hat{\mathbf{v}}$ its momentum, and $\hat{\mathbf{a}}$ its acceleration. The inverse inertia of a rigid body maps its momentum to its velocity.

The concept of inertia is generalised to include articulated-body inertias. An articulated body is an isolated system of rigid bodies which interacts with its environment through certain of its members (usually only one), which are called handles. An articulated-body inertia relates an external force applied to a handle to its acceleration, or, in the case of a cross-inertia, to the acceleration of a different handle. Articulated-body inertias are used to simplify rigid-body systems by reducing the apparent number of members.

This concludes the description of the spatial notation and underlying spatial algebra. In the rest of this book it will be used as a tool for analysing robot dynamics and expressing dynamics calculation algorithms. These two chapters have covered all the material necessary for the analyses in the remainder of this book.

# Chapter 4
# Inverse Dynamics -- The Recursive Newton-Euler Method

## 4.1. Introduction

Inverse dynamics is the calculation of the forces required at a robot's joints in order to produce a given set of joint accelerations. The principal uses of inverse dynamics are in robot control and trajectory planning. In control applications it is usually incorporated as an element in a feedback loop to convert accelerations computed according to some control law into the joint forces which will achieve those accelerations (e.g., see [27], [39], [33]). Rapid execution is essential to achieve real-time control. In trajectory planning, inverse dynamics can be used to check that the forces needed to execute a proposed trajectory do not exceed the actuators' limits [16], and recently discovered time-scaling properties of inverse dynamics have facilitated its use in planning minimum-time trajectories [52], [23]. Inverse dynamics is also used as a building block for constructing forward dynamics algorithms (see Chapter 5).

There have been two main approaches to formulating robot dynamics: the Lagrange approach and the Newton-Euler approach. They are equivalent in that they solve the same problem, but the route to the solution is different for each. In both approaches the key to computational efficiency is the use of recursive formulations, which not only reduce the computational requirement substantially over non-recursive formulations, but also reduce the computational complexity from (typically) $O(n^4)$ to $O(n)$.

In the Lagrange approach, the Lagrangian of the manipulator is expressed in terms of joint variables and velocities (which are the generalised coordinates and their derivatives), and the expression is substituted into the Euler-Lagrange equation which is expanded by symbolic differentiation to give the generalised forces (joint forces) in terms of the generalised coordinates, velocities and accelerations. This method was first applied to robot dynamics by Kahn [30], who adapted Uicker's Lagrangian formulation for more general mechanisms [66]. Kahn's formulation is computationally inefficient, but the efficiency of the more recently developed recursive Lagrangian formulation [22] is comparable with that of the recursive Newton-Euler formulation.

In the Newton-Euler approach, the equations of motion (Newton's + Euler's equations) are applied to each link and the resulting equations are combined with constraint equations from the joints in such a way as to give the joint forces in terms of the joint accelerations. This method was originally developed for multi-body satellite dynamics [26], [25], and has been applied to robot mechanisms by several authors [60], [71], [73], and [43]. Some of these formulations are partially recursive, but it was Luh, et al. [40] who developed the fully recursive Newton-Euler formulation.

Other relevant contributions include [2], [6], [27], [29], [32], [31], [34], [41], [47], [53], [54], [57], [64], [68], [74], [75], [76], and several important papers are reproduced in [10].

Whichever approach is used, a non-recursive method usually results in a set of equations of motion in the form

$$Q_i = \sum_{j=1}^{n} H_{ij}\ddot{q}_j + \sum_{j=1}^{n}\sum_{k=1}^{n} C_{ijk}\dot{q}_j\dot{q}_k + g_i , \tag{4.1}$$

where $Q_i$, $\dot{q}_i$, and $\ddot{q}_i$ are the joint forces, velocities and accelerations, and $H_{ij}$, $C_{ijk}$ and $g_i$ are inertial, Coriolis and gravitational coefficients, which are state dependent. Such an equation can be used directly for inverse dynamics, or it can be adapted for forward dynamics by solving for $\ddot{q}_i$. The main source of inefficiency is the explicit calculation of O $(n^2)$ quantities $H_{ij}$ and O $(n^3)$ quantities $C_{ijk}$. Recursive formulations avoid the need to calculate these quantities by expressing the equations of motion (implicitly) in terms of recurrence relations.

The recursive Newton-Euler formulation is the most efficient currently known general method for calculating inverse dynamics. However, a number of specialised algorithms have been proposed to calculate the inverse dynamics of particular kinds of robot more efficiently (e.g., see [24], [32], [47], [72]). Typically such algorithms take advantage of special properties of particular robot geometries and make simplifying assumptions about the inertia parameters of the links. The most successful algorithms are 2 to 3 times faster than the general Newton-Euler formulation applied to the same robot, but some less successful algorithms are actually slower than the general method. In principle, methods which use optimised symbolic expressions ought to produce faster algorithms than numerical methods, but this is not always the case in practice. Such algorithms will not be considered in this book.

The next section introduces recurrence relations and demonstrates how they can be used to generate efficient algorithms. The following section defines the system model which is used to describe the robot being analysed; the section after that derives the equations of the Newton-Euler formulation in absolute coordinates; and the final section refers these equations to link coordinates, where they may be evaluated most efficiently.

## 4.2. Recurrence Relations

A recurrence relation is an equation defining a member of a sequence in terms of its predecessors. Usually it is possible to generate any member of a sequence by the appropriate number of iterations of the recurrence relation, given suitable starting values, in which case the sequence is defined recursively by the recurrence relation and its starting values.

In the context of robot dynamics, the elements of the sequences would be, for example, the spatial velocities of the robot's links, or their accelerations, and a recurrence relation giving the velocity or acceleration of link $i$ in terms of that of link $i-1$ could be used to calculate the velocity or acceleration of each link in turn, given appropriate values for the first link.

It turns out that the proper use of recurrence relations leads to algorithms which are substantially faster than non-recursive methods, for reasons described in Chapter 1. For example, the recursive Newton-Euler method is approximately 100 times faster than the Uicker/Kahn method for a six-degree-of-freedom robot [22].

In view of the dramatic effect that recurrence relations have on the efficiency of robot dynamics computations, it is worthwhile spending a few moments discussing how they work. Basically, recurrence relations offer a way of minimising the amount of unnecessary calculation in evaluating certain kinds of mathematical expression. To see this, let us consider a simple example.

Let the matrix $\mathbf{B}$ be defined as the product of $n$ matrices $\mathbf{A}_i$ thus:

$$\mathbf{B} = \prod_{i=1}^{n} \mathbf{A}_i \,. \tag{4.2}$$

The task is to compute the numeric value of $\dot{\mathbf{B}}$ $(= \frac{d}{dt}\mathbf{B})$ given numeric values for $\mathbf{A}_i$ and $\dot{\mathbf{A}}_i$. Symbolic differentiation of Equation (4.2) gives the following equation for $\dot{\mathbf{B}}$ in terms of $\mathbf{A}_i$ and $\dot{\mathbf{A}}_i$:

$$\dot{\mathbf{B}} = \sum_{i=1}^{n} \mathbf{P}_i \quad \text{where} \quad \mathbf{P}_i = (\prod_{j=1}^{i-1} \mathbf{A}_j)\,\dot{\mathbf{A}}_i\,(\prod_{j=i+1}^{n} \mathbf{A}_j)\,. \tag{4.3}$$

To calculate $\dot{\mathbf{B}}$ directly from Equation (4.3) requires the calculation of $n$ matrices $\mathbf{P}_i$, each requiring $n-1$ multiplications, and their summation, requiring $n-1$ additions. The total amount of computation is $n(n-1)$ matrix multiplications and $n-1$ matrix additions, or $n(n-1)m + (n-1)a$ for short.

This is the 'algebraic equation' approach to computing the answer, and is characteristic of many early robot dynamics calculation schemes. The problem is solved algebraically giving a symbolic equation, and the numeric solution is obtained by substituting numeric values into the symbolic equation and evaluating it.

This approach is usually highly inefficient because of the large amount of repeated calculation in evaluating the solution. The repetition is not particularly obvious, but is brought about by the fact that an algebraic solution to a problem like this contains a great many terms which are almost identical but which are each calculated independently of the others. In this case, $\mathbf{P}_i$ differs from $\mathbf{P}_j$ by only 2 factors out of $n$, but each is calculated independently. For example, consider the case when $n=5$. The computation performed is the following:

$$\begin{aligned}\dot{\mathbf{B}} = \; & \dot{\mathbf{A}}_1\,\mathbf{A}_2\,\mathbf{A}_3\,\mathbf{A}_4\,\mathbf{A}_5 + \\ & \mathbf{A}_1\,\dot{\mathbf{A}}_2\,\mathbf{A}_3\,\mathbf{A}_4\,\mathbf{A}_5 + \\ & \mathbf{A}_1\,\mathbf{A}_2\,\dot{\mathbf{A}}_3\,\mathbf{A}_4\,\mathbf{A}_5 + \\ & \mathbf{A}_1\,\mathbf{A}_2\,\mathbf{A}_3\,\dot{\mathbf{A}}_4\,\mathbf{A}_5 + \\ & \mathbf{A}_1\,\mathbf{A}_2\,\mathbf{A}_3\,\mathbf{A}_4\,\dot{\mathbf{A}}_5\,.\end{aligned}$$

Notice that $\mathbf{A}_1\,\mathbf{A}_2$ is calculated three times, and in two of those cases the result is multiplied by $\mathbf{A}_3$. Similar remarks apply to $\mathbf{A}_3$, $\mathbf{A}_4$ and $\mathbf{A}_5$. This repetition escalates as $n$ increases.

Calculating the answer by means of recurrence relations offers a way of eliminating these unnecessary calculations. To solve the problem using recurrence relations, introduce the sequence of partial results $\mathbf{B}_i = \prod_{j=1}^{i} \mathbf{A}_j$, in which case $\mathbf{B}_{i+1}$ is given in terms of $\mathbf{B}_i$ by the recurrence relation

$$\mathbf{B}_{i+1} = \mathbf{B}_i \, \mathbf{A}_{i+1} \, . \tag{4.4}$$

Repeated application of Equation (4.4), starting with $\mathbf{B}_1 = \mathbf{A}_1$, allows us to calculate all the $\mathbf{B}_i$, and hence $\mathbf{B}$; but the task is to calculate $\dot{\mathbf{B}}$, not $\mathbf{B}$, so we differentiate Equation (4.4) to get a recurrence relation in $\dot{\mathbf{B}}_i$:

$$\dot{\mathbf{B}}_{i+1} = \dot{\mathbf{B}}_i \, \mathbf{A}_{i+1} + \mathbf{B}_i \, \dot{\mathbf{A}}_{i+1} \, . \tag{4.5}$$

Now we can calculate $\dot{\mathbf{B}}$ from Equation (4.5), starting with $\dot{\mathbf{B}}_1 = \dot{\mathbf{A}}_1$. To calculate $\mathbf{B}$ this way requires $n-1$ iterations of Equation (4.5), where one iteration requires $2m+1a$, but we also need the values of $\mathbf{B}_1 \ldots \mathbf{B}_{n-1}$, which requires $n-2$ iterations of Equation (4.4) (or none if $n < 2$). So the overall computational requirement is $(3n-4)m + (n-1)a$.

| size $n$ | algebraic $n^2-n$ | recurrence relation $3n-4$ |
|---|---|---|
| 1 | 0 | 0 |
| 2 | 2 | 2 |
| 3 | 6 | 5 |
| 5 | 20 | 11 |
| 10 | 90 | 26 |
| 20 | 380 | 56 |

**Table 4-1:** Comparison of the computational requirements of the algebraic and recurrence relation methods

Table 4-1 shows the computational requirements (for multiplication only) of the two methods for various values of $n$. Notice that the computational complexity of the recurrence relation method is lower than that of the algebraic equation method (O $(n)$ vs. O $(n^2)$), so the improvement increases with problem size. The degree of improvement is greater with a more complicated problem, like robot dynamics, where the computational complexity is improved from O $(n^4)$ to O $(n)$.

## 4.3. The Robot System Model

The dynamics algorithms in this and the following chapters are described using a system model of the robot. This is a description of the robot in terms of the number of links, their inertias, and the joints connecting them.

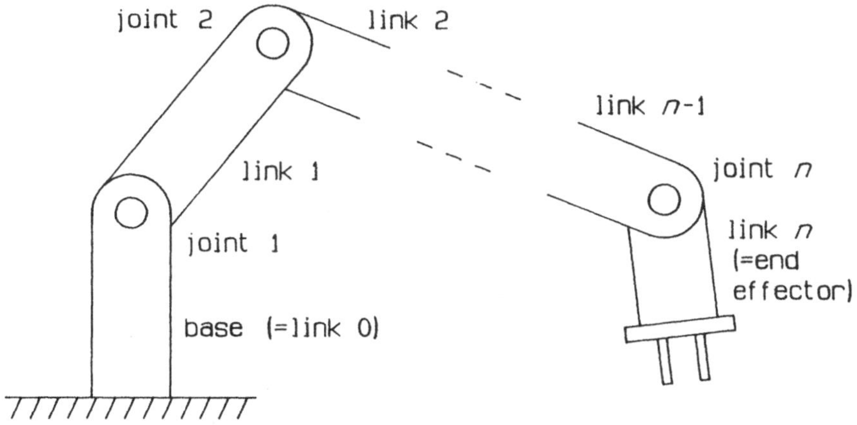


**Figure 4-1:** Robot connectivity

The first step in defining a system model is to label the various components (joints and links) and specify which is connected to which. Normally the connectivity[9] has to be given explicitly, but for the moment we will consider only robots consisting of a single un-branched open-loop kinematic chain, as shown in Figure 4-1, for which the connectivity is

---

[9]Connectivity is used here in the topological sense.

implicit in the numbering scheme. More general robot mechanisms will be considered in Chapters 7 and 9.

Let there be n movable links, numbered $1..n$, and $n$ one-degree-of-freedom joints, also numbered $1..n$, such that joint $i$ connects link $i$ to link $i-1$. Joint 1 connects link 1 to the immovable base member, which will be called link 0 for convenience.

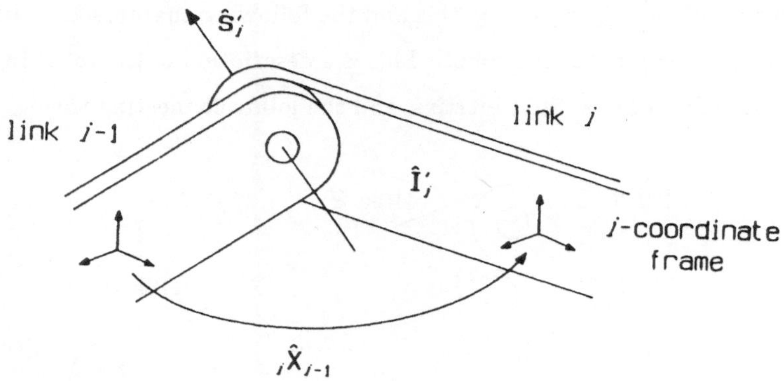


**Figure 4-2:** Robot geometry and parameters

The next step is to define the geometry of the linkage, the types of joint and the inertia parameters (see Figure 4-2). To do this, a set of link coordinate systems is introduced, one attached to each link, where the coordinate system attached to link $i$ is called $i$-coordinates. There is no restriction on where these coordinate systems are placed relative to their links, but certain placements allow the dynamics to be computed more efficiently than others. This topic is discussed in Chapter 8. Since the position of a link is fixed in its own coordinate system, we can express the inertia of link $i$ and the axis of any joint embedded in link $i$ as a constant in $i$-coordinates. Let $\hat{\mathbf{I}}'_i$ be the spatial inertia of link $i$ and let $\hat{\mathbf{s}}'_i$ describe the motion allowed by joint $i$, both expressed in $i$-coordinates. $\hat{\mathbf{s}}'_i$ is either a unit line vector or a unit free vector, depending on whether joint $i$ is revolute or prismatic. (Other kinds of joint are considered in Chapter 7.)

Having introduced link coordinates, we need to be able to transform representations of objects into and out of link coordinates. To do this, the adjacent-link coordinate transformations ${}_i\hat{\mathbf{X}}_{i-1}$ are introduced. Each such transformation is a function of one joint variable, and needs to be computed on demand. The details of how this is done depend on the nature of the joint and how the link coordinate systems are chosen, and are left to Chapter 8. All other coordinate transformations can be computed from the adjacent-link transformations. It will be convenient at this stage to make 0-coordinates synonymous with absolute coordinates.

Finally, we need to define the system variables. These are as follows: $q_i$ is the joint variable for joint $i$, $\dot{q}_i$ its velocity, $\ddot{q}_i$ its acceleration and $Q_i$ the joint force. They may be treated as the components of the $n$-dimensional vectors $\mathbf{q}$, $\dot{\mathbf{q}}$, $\ddot{\mathbf{q}}$ and $\mathbf{Q}$. We shall always assume that $\mathbf{q}$ and $\dot{\mathbf{q}}$ are known and that we are either trying to find $\ddot{\mathbf{q}}$ given $\mathbf{Q}$ (forward dynamics) or $\mathbf{Q}$ given $\ddot{\mathbf{q}}$ (inverse dynamics) at some particular instant.

To simplify the task of describing the dynamics algorithms, they will each be derived initially in absolute coordinates. To facilitate this the quantities $\hat{\mathbf{s}}_i$ and $\hat{\mathbf{I}}_i$ are introduced, which are the joint motion and spatial inertia of link $i$ expressed in absolute coordinates. Together they define the instantaneous kinematics and inertia of the system. They may be calculated at any instant from the formulae

$$\hat{\mathbf{s}}_i = {}_0\hat{\mathbf{X}}_i\,\hat{\mathbf{s}}_i' \quad \text{and} \quad \hat{\mathbf{I}}_i = {}_0\hat{\mathbf{X}}_i\,\hat{\mathbf{I}}_i'\,{}_i\hat{\mathbf{X}}_0\,,$$

where the transformation ${}_0\hat{\mathbf{X}}_i$ is calculated from the recurrence relation

$${}_0\hat{\mathbf{X}}_i = {}_0\hat{\mathbf{X}}_{i-1}\,{}_{i-1}\hat{\mathbf{X}}_i\,.$$

This model has been kept simple to facilitate the description of the basic dynamics algorithms, and will be generalised in Chapter 7. A system model for arbitrary rigid-body systems (which includes robots) is described in [76].

## 4.4. The Recursive Newton-Euler Formulation

The problem is stated as follows: given a system model of a robot as outlined in the previous section and the values of the desired joint accelerations, find the joint forces required to produce those accelerations. The solution to this problem by the recursive Newton-Euler method consists of three steps:

1. calculate the velocity and acceleration of each link,

2. calculate the net force acting on each link from its motion and inertia,

3. calculate the joint forces required to produce the forces in step 2.

### Step 1: Link Motion

Let $\hat{\mathbf{v}}_i$ be the absolute velocity of link $i$ and $\hat{\mathbf{a}}_i$ its absolute acceleration. The velocity of link $i$ can be expressed as the sum of the velocity of link $i-1$ (the preceding link) and the velocity across joint $i$, which is the relative velocity of link $i$ with respect to link $i-1$. This gives rise to a recurrence relation

$$\hat{\mathbf{v}}_i = \hat{\mathbf{v}}_{i-1} + \hat{\mathbf{s}}_i \dot{q}_i . \qquad (\hat{\mathbf{v}}_0 = \hat{\mathbf{0}}) \tag{4.6}$$

(The starting value is shown in brackets.) The recurrence relation for accelerations is obtained by differentiating Equation (4.6) and is given by

$$\hat{\mathbf{a}}_i = \hat{\mathbf{a}}_{i-1} + \hat{\mathbf{v}}_i \times \hat{\mathbf{s}}_i \dot{q}_i + \hat{\mathbf{s}}_i \ddot{q}_i . \qquad (\hat{\mathbf{a}}_0 = \hat{\mathbf{0}}) \tag{4.7}$$

Successive iterations of Equations (4.6 and 4.7) allow us to calculate the velocity and acceleration of every link.

### Step 2: Net Forces on the Links

The net force acting on link $i$ is given by the spatial equation of motion as the link's rate of change of momentum:

$$\hat{\mathbf{f}}_i^* = \frac{d}{dt}(\hat{\mathbf{I}}_i \hat{\mathbf{v}}_i)$$

$$= \hat{\mathbf{I}}_i \hat{\mathbf{a}}_i + \hat{\mathbf{v}}_i \hat{\times} \hat{\mathbf{I}}_i \hat{\mathbf{v}}_i . \qquad (4.8)$$

**Step 3: Joint Forces**

First we must find the total force transmitted from link $i-1$ to link $i$ through joint $i$, which includes the component due to bearing reaction forces. If $\hat{\mathbf{f}}_i$ is this force, then we have that

$$\hat{\mathbf{f}}_i - \hat{\mathbf{f}}_{i+1} = \hat{\mathbf{f}}_i^* ,$$

which can be rearranged to give a recurrence relation

$$\hat{\mathbf{f}}_i = \hat{\mathbf{f}}_{i+1} + \hat{\mathbf{f}}_i^* . \quad (\hat{\mathbf{f}}_n = \hat{\mathbf{f}}_n^*) \qquad (4.9)$$

This recurrence relation allows us to calculate all the spatial joint forces starting from the end effector and working towards the base. Known external forces acting on individual links can be included in Equation (4.9): if $\hat{\mathbf{f}}_i^x$ is the external force acting on link $i$ then

$$\hat{\mathbf{f}}_i = \hat{\mathbf{f}}_{i+1} + \hat{\mathbf{f}}_i^* - \hat{\mathbf{f}}_i^x$$

(see Figure 4-3). Gravity may be dealt with in this manner, but it is more efficient to simulate gravity by giving the robot's base a fictitious acceleration. This can be done by using $\hat{\mathbf{a}}_0 = -\hat{\mathbf{g}}$ as the starting value for Equation (4.7), where $\hat{\mathbf{g}}$ is the gravitational acceleration vector.

The scalar joint forces are the components of the spatial joint forces which do work in the direction of joint motion (i.e., the active components of these forces) and are given by[10]

$$Q_i = \hat{\mathbf{s}}_i^S \hat{\mathbf{f}}_i . \qquad (4.10)$$

---

[10] This equation is derived from the power balance equation, which in spatial form is $\hat{\mathbf{v}}^S \hat{\mathbf{f}} = \dot{q} Q$.

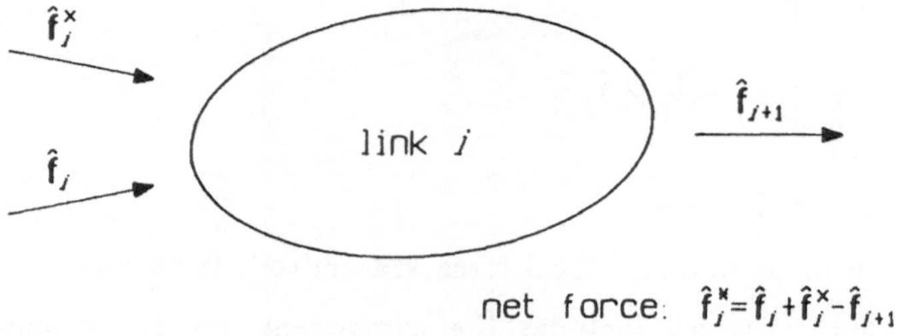


**Figure 4-3:** Forces acting on a link

Equations (4.6-4.10) form the complete Newton-Euler formulation for inverse dynamics, expressed in absolute coordinates. The calculation scheme is first to calculate $\hat{\mathbf{s}}_i$ and $\hat{\mathbf{I}}_i$ for each link as described in Section 4.3, then to perform the calculations indicated in Equations (4.6-4.10). $n-1$ iterations of Equation (4.9) are required and $n$ iterations of each of the others.

## 4.5. Reformulating the Equations in Link Coordinates

A more efficient calculation scheme can be obtained by performing those calculations pertinent to link $i$ in $i$-coordinates. The increased efficiency is due mainly to it being easier to transform three vector quantities ($\hat{\mathbf{v}}_i$, $\hat{\mathbf{a}}_i$ and $\hat{\mathbf{f}}_i$) from one link coordinate system to an adjacent one than it is to transform one vector and a tensor ($\hat{\mathbf{s}}_i'$ and $\hat{\mathbf{I}}_i'$) from link coordinates to absolute coordinates. Further computational savings can be made by placing link coordinate systems so that the vectors $\hat{\mathbf{s}}_i'$ are aligned with a link coordinate axis, allowing short cuts to be taken in evaluating expressions involving $\hat{\mathbf{s}}_i'$. This topic is discussed in Chapter 8.

Converting a formulation in absolute coordinates to one in link coordinates is very simple. Firstly, all the quantities defined as being represented in absolute coordinates are replaced by or redefined as quantities represented in link coordinates. So $\hat{\mathbf{I}}_i$ is replaced by $\hat{\mathbf{I}}_i'$, $\hat{\mathbf{v}}_i$ is redefined as the velocity of link $i$ represented in $i$-coordinates, and so on. Secondly, quantities on the right-hand side of an equation which are represented in a coordinate system other than that of the quantity on the left-hand side are transformed accordingly.

The result of applying these rules to Equations (4.6-4.10) is as follows:

$$\hat{\mathbf{v}}_i = {}_i\hat{\mathbf{X}}_{i-1}\,\hat{\mathbf{v}}_{i-1} + \hat{\mathbf{s}}_i'\,\dot{q}_i\,, \qquad (\hat{\mathbf{v}}_0 = \hat{\mathbf{0}}) \tag{4.11}$$

$$\hat{\mathbf{a}}_i = {}_i\hat{\mathbf{X}}_{i-1}\,\hat{\mathbf{a}}_{i-1} + \hat{\mathbf{v}}_i \hat{\times} \hat{\mathbf{s}}_i'\,\dot{q}_i + \hat{\mathbf{s}}_i'\,\ddot{q}_i\,, \quad (\hat{\mathbf{a}}_0 = \hat{\mathbf{0}}) \tag{4.12}$$

$$\hat{\mathbf{f}}_i^* = \hat{\mathbf{I}}_i'\,\hat{\mathbf{a}}_i + \hat{\mathbf{v}}_i \hat{\times} \hat{\mathbf{I}}_i'\,\hat{\mathbf{v}}_i\,, \tag{4.13}$$

$$\hat{\mathbf{f}}_i = {}_i\hat{\mathbf{X}}_{i+1}\,\hat{\mathbf{f}}_{i+1} + \hat{\mathbf{f}}_i^*\,, \qquad (\hat{\mathbf{f}}_n = \hat{\mathbf{f}}_n^*) \tag{4.14}$$

$$Q_i = \hat{\mathbf{s}}_i'^S\,\hat{\mathbf{f}}_i\,. \tag{4.15}$$

# Chapter 5
# Forward Dynamics -- The Composite-Rigid-Body Method

## 5.1. Introduction

Forward (or direct) dynamics is the calculation of the acceleration of the robot's mechanism in response to the applied forces. It is used primarily for simulation, so it is not necessary for forward dynamics to meet the stringent speed requirements of real-time control applications. However, computational efficiency is still important to minimise the cost of simulation. The composite-rigid-body method (or Walker-Orin method) was originally described in [74] as method 3, and is the most efficient method available for calculating forward dynamics in most practical cases; although there are situations where the articulated-body method described in Chapter 6 is more efficient. The composite-rigid-body method has been used in the simulators described in [19], [20], [61]. Some other approaches to dynamics simulation are described in [1], [3], [6], [31], [32], [34], [55], [62].

Forward dynamics is a more difficult problem to solve than inverse dynamics, and this is reflected in the fact that the computational requirements of forward dynamics algorithms are higher than those of inverse dynamics algorithms. With the inverse dynamics problem we are given enough information to be able to calculate the motion of every link in the mechanism, and it is then an easy matter to calculate the forces involved. With the forward dynamics problem we do not know in advance the motion of the system (the joint accelerations are unknown), and we do not know enough about the forces to be able to calculate directly the forces acting on the links.

There are basically two approaches to solving the forward dynamics problem:

- calculate recursion coefficients which propagate motion and force constraints along the mechanism allowing the problem to be solved directly, or

- obtain and then solve a set of simultaneous equations in the unknown joint accelerations.

Few algorithms adopt the first approach; but an example of one that does is the articulated-body method described in Chapter 6. Algorithms which use this approach are more difficult to derive, but they can achieve $O(n)$ computational complexity. The majority of published algorithms, including the composite-rigid-body method, adopt the second approach. Algorithms which use this approach can only achieve $O(n^3)$ computational complexity, since they involve solving a set of $n$ simultaneous equations, but they can be very efficient for small values of $n$. In particular, the composite-rigid-body method, which is the most efficient of this class of algorithm, is more efficient than the articulated-body method for $n < 9$, so it is the most efficient algorithm available for most practical problems.

The equations of motion for a robot mechanism can be written in the form

$$\mathbf{Q} = \mathbf{H}(\mathbf{q})\,\ddot{\mathbf{q}} + \mathbf{C}(\mathbf{q},\dot{\mathbf{q}})\,, \tag{5.1}$$

which is a vectorial rendering of Equation (4.1). $\mathbf{H}$ is an $n \times n$ matrix composed of the inertia coefficients $H_{ij}$, and will be referred to as the joint-space inertia matrix. The elements of $\mathbf{C}$ encompass the remaining terms in Equation (4.1), and $\mathbf{q}$ , $\dot{\mathbf{q}}$ , $\ddot{\mathbf{q}}$ and $\mathbf{Q}$ are the vectors of generalised (or joint-space) coordinates, velocities, accelerations and forces. The elements of $\mathbf{H}$ and $\mathbf{C}$ are independent of the unknown accelerations, and can be calculated in a straightforward manner. The result is a system of $n$ simultaneous linear equations in $\ddot{\mathbf{q}}$. $\mathbf{H}$ is symmetric and positive definite,[11] and this fact can be used to speed up the forward dynamics calculation.

The process can be broken down into three steps:

1. calculate the elements of $\mathbf{H}$

2. calculate the elements of $\mathbf{C}$

3. solve the set of simultaneous equations $\mathbf{H}\ddot{\mathbf{q}} = \mathbf{Q} - \mathbf{C}$.

There are many standard methods for accomplishing the third step, but in view of the properties of $\mathbf{H}$ a method specific to symmetric, positive-definite matrices may be more efficient. Most non-recursive algorithms calculate the elements of $\mathbf{C}$ via the multiple sum given in Equation (4.1), but it is shown in the next section how this can be done more efficiently using a recursive inverse dynamics algorithm. Calculating the elements of $\mathbf{H}$ can also be accomplished using an inverse dynamics algorithm, but it is more efficient to use the composite-rigid-body method, which is described in Section 5.3. Note that, strictly speaking, the composite-rigid-body method is an algorithm for calculating $\mathbf{H}$ only, and does not address the other parts of the forward dynamics calculation.

## 5.2. Using an Inverse Dynamics Algorithm to Calculate Forward Dynamics

As mentioned in Chapter 4, most non-recursive dynamics algorithms explicitly calculate the elements of $\mathbf{H}$ and $\mathbf{C}$, and are therefore easily adapted to calculate forward dynamics, but recursive algorithms do not. However, it is possible to calculate both $\mathbf{H}$ and $\mathbf{C}$ using a recursive inverse dynamics algorithm. With an $O(n)$ algorithm the calculation of $\mathbf{H}$ can be

[11] The simplest way to show this is to express the kinetic energy of the mechanism as a quadratic form $KE = 1/2\, \dot{\mathbf{q}}^T \mathbf{A}\, \dot{\mathbf{q}}$, where $\mathbf{A}$ is required to be symmetric and positive definite. Substituting this into Lagrange's equation gives, ultimately, $\mathbf{H} = \mathbf{A}$.

achieved in $O(n^2)$ steps and $\mathbf{C}$ in $O(n)$. These are the minimum possible computational complexities, since $\mathbf{H}$ and $\mathbf{C}$ contain $n^2$ and $n$ elements respectively. This approach produces a substantial increase in efficiency over non-recursive methods, particularly with the calculation of $\mathbf{C}$, but $\mathbf{H}$ can be calculated more quickly using the composite-rigid-body method.

**Calculating C**

If we set the generalised acceleration vector in Equation (5.1) to zero, then the equation simplifies to:

$$\mathbf{Q} = \mathbf{C}(\mathbf{q},\dot{\mathbf{q}})\,. \tag{5.2}$$

The vector $\mathbf{C}$ may therefore be interpreted as the vector of generalised forces which produces zero generalised acceleration, given the robot's current position and velocity. If we view the inverse dynamics algorithm as a function $IDA(\mathbf{q},\dot{\mathbf{q}},\ddot{\mathbf{q}})$, which takes the position, velocity and acceleration as arguments, then we may compute $\mathbf{C}$ by

$$\mathbf{C}(\mathbf{q},\dot{\mathbf{q}}) = IDA(\mathbf{q},\dot{\mathbf{q}},\mathbf{0})\,. \tag{5.3}$$

This is an efficient, $O(n)$ computation which compares very favourably with the equivalent $O(n^4)$ computation performed by a typical non-recursive algorithm. It is possible to formulate a specialised version of the inverse dynamics algorithm with the assumption $\ddot{\mathbf{q}}=\mathbf{0}$ built in, but the computational saving is negligible.

**Calculating H**

From Equations (5.1 and 5.3) and the definition of the inverse dynamics function we have that

$$IDA(\mathbf{q},\dot{\mathbf{q}},\ddot{\mathbf{q}}) = \mathbf{H}\ddot{\mathbf{q}} + \mathbf{C}$$

$$= \mathbf{H}\ddot{\mathbf{q}} + IDA(\mathbf{q},\dot{\mathbf{q}},\mathbf{0})\,,$$

so

$$\mathbf{H}\ddot{\mathbf{q}} = IDA(\mathbf{q},\dot{\mathbf{q}},\ddot{\mathbf{q}}) - IDA(\mathbf{q},\dot{\mathbf{q}},\mathbf{0}) . \tag{5.4}$$

Using Equation (5.4) we can calculate the product of **H** with any vector of our choice. However, Equation (5.4) as it stands is not a very efficient way of doing it since the velocity (and gravity) terms are calculated twice and then subtracted, when there is no need to calculate them at all. It is better to formulate a simplified version of the inverse dynamics algorithm, $IDA2(\mathbf{q},\ddot{\mathbf{q}})$, which ignores velocity effects, etc., and calculates only those terms which do not cancel, so that

$$\mathbf{H}\ddot{\mathbf{q}} = IDA2(\mathbf{q},\ddot{\mathbf{q}}) . \tag{5.5}$$

If we let $\ddot{\mathbf{q}} = \delta_i$, where $\delta_i$ is defined such that its $i^{th}$ element is unity and the rest are zero, then the result of calculating $IDA2(\mathbf{q},\delta_i)$ is simply column $i$ of **H**; so we can calculate the whole of **H** with $n$ applications of Equation (5.5), one per column. This is Walker and Orin's first method for calculating **H** [74]. Their second method is basically the same, but achieves a greater efficiency by noting that **H** is symmetric and arranging to calculate only the lower triangle.

### Calculating the Acceleration

Having calculated **C** and **H**, the final step is to solve the set of simultaneous equations

$$\mathbf{H}\ddot{\mathbf{q}} = \mathbf{Q} - \mathbf{C} . \tag{5.6}$$

The complexity of this step is $O(n^3)$, thus making the complexity of the overall solution $O(n^3)$, but the coefficient of $n^3$ in the computational requirement is much smaller than that of any of the lower powers of $n$, and for values of $n$ around $n$=6 (or even $n$=12) its contribution is insignificant. The largest part of the computational burden for small $n$ is the calculation of **H** if the above method is used, or of **C** if **H** is calculated by the composite-rigid-body method.

Since $\mathbf{H}$ is symmetric and positive definite, specialised techniques like the $L\,L^T$ or $L\,D\,L^T$ factorisations may solve the equations more efficiently. It is possible for $\mathbf{H}$ to become singular if some of the links are given zero (or singular) spatial inertias.

In a simulation context it is often the case that $\mathbf{Q}$ changes much more rapidly than $\mathbf{H}$ or $\mathbf{C}$. This will happen if, for example, the robot is being controlled by a stiff control system but isn't moving very quickly. In this situation $\mathbf{H}$ and $\mathbf{C}$ change very little from one integration time step to the next, and it makes sense to store the calculated values of $\mathbf{C}$ and $\mathbf{H}$ (after factorisation) for use over several time steps [19]. Some accuracy is lost, but a considerable amount of computation is saved.

Walker and Orin also describe a method for calculating $\ddot{\mathbf{q}}$ which bypasses the need to calculate $\mathbf{H}$ explicitly (method 4). This method uses an iterative technique for solving Equation (5.6) that converges to an exact solution in a maximum of $n$ iterations. At each iteration this method requires the value of a vector $\mathbf{H}\,\mathbf{u}$, where $\mathbf{u}$ is chosen by the iteration algorithm. The complexity of this method is $O(n^2)$, but the coefficient of $n^2$ is large enough that it ends up being less efficient that the composite-rigid-body method for all but very large values of $n$.

## 5.3. The Composite-Rigid-Body Method

This is a method for calculating the joint-space inertia matrix which is more efficient than any of the methods described in the previous section. It is described in [74] as method 3. The method works by treating links $i\,..\,n$ as a composite rigid body (hence the name) which is accelerating about joint $i$, and uses the inertia of the composite body to calculate the forces involved.

From Equation (5.5) and the discussion in the previous section, it is evident that $\mathbf{H}\,\ddot{\mathbf{q}}$ is the vector of joint forces required to impart a joint-space acceleration of $\ddot{\mathbf{q}}$ to a stationary robot which is free from gravitational or other external influences. In particular, column $i$ of $\mathbf{H}$ (which is $\mathbf{H}\,\delta_i$) is the vector of joint forces required to produce a unit acceleration about joint $i$ and zero acceleration about all other joints (see Figure 5-1). Since no motion occurs about any of the joints $i{+}1\ .\ .\ n$, they can all be replaced by rigid connections without having any effect on the dynamics lower down the chain; so the moving part of the robot simplifies to a single composite rigid body.

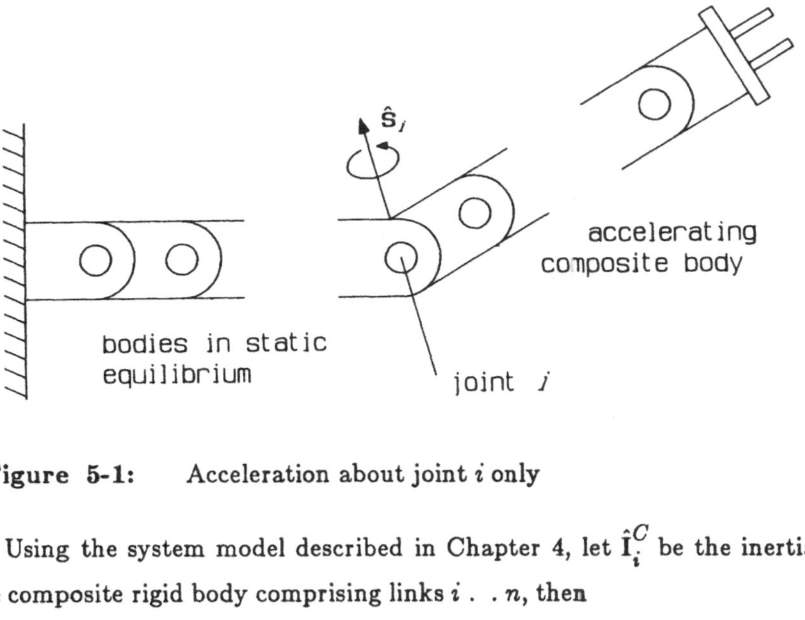


**Figure 5-1:** Acceleration about joint $i$ only

Using the system model described in Chapter 4, let $\hat{\mathbf{I}}_i^C$ be the inertia of the composite rigid body comprising links $i\ .\ .\ n$, then

$$\hat{\mathbf{I}}_i^C = \sum_{j=i}^{n} \hat{\mathbf{I}}_j\,,$$

which expressed as a recurrence relation is

$$\hat{\mathbf{I}}_i^C = \hat{\mathbf{I}}_{i+1}^C + \hat{\mathbf{I}}_i\,. \qquad (\,\hat{\mathbf{I}}_n^C = \hat{\mathbf{I}}_n\,) \tag{5.7}$$

The force required to give $\hat{\mathbf{I}}_i^C$ an acceleration of $\hat{\mathbf{s}}_i$ (i.e., unit acceleration about joint axis $i$) is

$$\hat{\mathbf{f}}_i = \hat{\mathbf{I}}_i^C \hat{\mathbf{s}}_i . \tag{5.8}$$

Now links 1 . . $i-1$ are in static equilibrium, so there is no net force acting on any of them. Therefore every joint from 1 to $i$ transmits $\hat{\mathbf{f}}_i$ on to the next link. So for each joint from 1 to $i$ we have that the scalar joint force is the component of $\hat{\mathbf{f}}_i$ on the appropriate joint axis, and these quantities are the first $i$ elements of column $i$ of $\mathbf{H}$. Thus we have:

$$H_{ji} = \hat{\mathbf{s}}_j^S \hat{\mathbf{f}}_i , \qquad (j \leq i) \tag{5.9}$$

which, making use of the symmetry of $\mathbf{H}$, enables us to calculate the whole of the matrix $\mathbf{H}$. An equation which shows the symmetry more clearly is

$$H_{ij} = H_{ji} = \hat{\mathbf{s}}_j^S \hat{\mathbf{I}}_r^C \hat{\mathbf{s}}_i , \qquad \text{where } r = \max(i,j) . \tag{5.10}$$

Equations (5.7-5.9) form the composite-rigid-body method for calculating the joint-space inertia matrix. To complete the forward dynamics algorithm we need to calculate $\mathbf{C}$ and to solve $\mathbf{H}\ddot{\mathbf{q}} = \mathbf{Q} - \mathbf{C}$, both of which were covered in the previous section.

## 5.4. The Composite-Rigid-Body Method in Link Coordinates

The advantage of calculating the joint-space inertia matrix in link coordinates is less clear than was the case with the inverse dynamics algorithm of Chapter 4, since in absolute coordinates we only need to transform the quantities $\hat{\mathbf{s}}_i'$ and $\hat{\mathbf{I}}_i'$ to serve for the calculation of both $\mathbf{C}$ and $\mathbf{H}$, whereas in link coordinates we need to transform $\hat{\mathbf{f}}_i$ and $\hat{\mathbf{I}}_i^C$ (the former involving $O(n^2)$ transformations) in addition to the transformations carried out by the inverse dynamics algorithm. Nevertheless, the link coordinate version is faster for $n \leq 16$ (see Chapter 8).

The link coordinate version of the algorithm is obtained as described in

Chapter 4, except that since $\hat{\mathbf{f}}_i$ is needed in several coordinate systems we replace it with $\hat{\mathbf{f}}_{ij}$ where the second subscript indicates the coordinate system in which it is represented. The link coordinate version is as follows:

$$\hat{\mathbf{I}}_i^C = {}_i\hat{\mathbf{X}}_{i+1}\,\hat{\mathbf{I}}_{i+1}^C\,{}_{i+1}\hat{\mathbf{X}}_i + \hat{\mathbf{I}}_i', \qquad (\hat{\mathbf{I}}_n^C = \hat{\mathbf{I}}_n') \tag{5.11}$$

$$\hat{\mathbf{f}}_{ii} = \hat{\mathbf{I}}_i^C\,\hat{\mathbf{s}}_i', \tag{5.12}$$

$$\hat{\mathbf{f}}_{ij} = {}_j\hat{\mathbf{X}}_{j+1}\,\hat{\mathbf{f}}_{ij+1}, \qquad (j < i) \tag{5.13}$$

$$H_{ij} = H_{ji} = \hat{\mathbf{s}}_j'^S\,\hat{\mathbf{f}}_{ij}. \qquad (j \le i) \tag{5.14}$$

## 5.5. The Lagrange Approach to the Composite-Rigid-Body Method

Just as the equations of the Newton-Euler formulation for inverse dynamics can be arrived at via Lagrange's equation, so too can the equations of the composite-rigid-body method.

From Equation (4.1) it follows that

$$H_{ij} = \frac{\partial Q_i}{\partial \ddot{q}_j}. \tag{5.15}$$

Lagrange's equation for $Q_i$ is

$$Q_i = \frac{d}{dt}\frac{\partial L}{\partial \dot{q}_i} - \frac{\partial L}{\partial q_i}, \tag{5.16}$$

where the Lagrangian is given by

$$L = KE - PE$$

$$= \sum_{k=1}^{n} 1/2\,\hat{\mathbf{v}}_k^S\,\hat{\mathbf{I}}_k\,\hat{\mathbf{v}}_k - PE. \tag{5.17}$$

Substituting Equations (5.16 and 5.17) into Equation (5.15) gives

$$H_{ij} = \frac{\partial}{\partial \ddot{q}_j} \left( \frac{d}{dt} \frac{\partial}{\partial \dot{q}_i} \left( \sum_{k=1}^{n} 1/2\, \hat{\mathbf{v}}_k^S \hat{\mathbf{I}}_k \hat{\mathbf{v}}_k - PE \right) - \frac{\partial L}{\partial q_i} \right). \tag{5.18}$$

Now $PE$ is independent of $\dot{\mathbf{q}}$ and $\frac{\partial L}{\partial q_i}$ is independent of $\ddot{\mathbf{q}}$, so Equation (5.18) simplifies to

$$H_{ij} = \frac{\partial}{\partial \ddot{q}_j} \frac{d}{dt} \frac{\partial}{\partial \dot{q}_i} \sum_{k=1}^{n} 1/2\, \hat{\mathbf{v}}_k^S \hat{\mathbf{I}}_k \hat{\mathbf{v}}_k . \tag{5.19}$$

$\hat{\mathbf{v}}_k$ is given in terms of joint velocities by

$$\hat{\mathbf{v}}_k = \sum_{i=1}^{k} \hat{\mathbf{s}}_i \dot{q}_i ,$$

so

$$\frac{\partial}{\partial \dot{q}_i} \hat{\mathbf{v}}_k = \begin{cases} \hat{\mathbf{0}} & \text{if } i > k \\ \hat{\mathbf{s}}_i & \text{if } i \le k \end{cases} .$$

A similar equation applies for $\frac{\partial}{\partial \ddot{q}_j} \hat{\mathbf{a}}_k$. Applying the differentiation operators, Equation (5.19) simplifies to

$$H_{ij} = \frac{\partial}{\partial \ddot{q}_j} \frac{d}{dt} \sum_{k=i}^{n} \hat{\mathbf{s}}_i^S \hat{\mathbf{I}}_k \hat{\mathbf{v}}_k ,$$

$$= \frac{\partial}{\partial \ddot{q}_j} \sum_{k=i}^{n} ((\hat{\mathbf{v}}_i \hat{\times} \hat{\mathbf{s}}_i)^S \hat{\mathbf{I}}_k \hat{\mathbf{v}}_k + \hat{\mathbf{s}}_i^S (\hat{\mathbf{I}}_k \hat{\mathbf{a}}_k + \hat{\mathbf{v}}_k \hat{\times} \hat{\mathbf{I}}_k \hat{\mathbf{v}}_k))$$

$$= \sum_{k=r}^{n} \hat{\mathbf{s}}_i^S \hat{\mathbf{I}}_k \hat{\mathbf{s}}_j \quad (r = \max(i,j))$$

$$= \hat{\mathbf{s}}_i^S \hat{\mathbf{I}}_r^C \hat{\mathbf{s}}_j . \tag{5.20}$$

Equation (5.20) is identical to Equation (5.10).

# Chapter 6
# Forward Dynamics -- The Articulated-Body Method

## 6.1. Introduction

This chapter describes the articulated-body method for forward dynamics -- an alternative to the composite-rigid-body method. This method has a computational complexity of O $(n)$ as compared with O $(n^3)$ for the composite-rigid-body method, although it does not become faster than the composite-rigid-body method until $n$ reaches a value of approximately 9.

The articulated-body method uses a different approach to solving forward dynamics from the methods described in the previous chapter. The basic idea is to regard the robot as consisting of a base member (whose motion is known), a single joint, and a single moving link which is in fact an articulated body representing the rest of the robot. The forward dynamics problem for this one-joint robot is easily solved once the apparent inertia of the moving link is known. Having found the acceleration of the first joint, the articulated body itself can be treated as a robot and the same process applied to obtain the acceleration of the next joint, and so on. So the articulated-body method consists of the calculation of a series of articulated-body inertias which are used to solve the forward dynamics one joint at a time.

The articulated-body method is an example of an algorithm which solves forward dynamics by the constraint propagation approach. Other examples are described by Armstrong [2], and Vereshchagin [68]. Armstrong's method also achieves $O(n)$ complexity, and uses recursion coefficients playing a similar role to articulated-body inertias, but the approach is somewhat different and in its basic form is only applicable to robots with spherical joints. Vereshchagin's algorithm is practically identical to the articulated-body method, but it is derived in a very different manner: applying directly Gauss' principle of least constraint and solving the resulting minimisation problem using a dynamic programming recursion relation.

An early version of the articulated-body method is described in [21]. The main difference between the versions described here and in that paper is that the version described here solves the forward dynamics problem in its entirety, whereas the version described in the paper neglects velocity dependent terms and assumes that the forces required to compensate for them have been calculated by an inverse dynamics algorithm.

### 6.1.1. Articulated-Body Inertias

Articulated-body inertias were described in Chapter 3, so only the important points will be mentioned here.

A collection of rigid bodies connected by joints is called an articulated body. To define an articulated-body inertia, we single out a particular member of the articulated body, called the handle, and define the articulated-body inertia as the relationship between a test force $\hat{\mathbf{f}}$ applied to the handle and the acceleration $\hat{\mathbf{a}}$ of the handle according to

$$\hat{\mathbf{f}} = \hat{\mathbf{I}}^A \hat{\mathbf{a}} + \hat{\mathbf{p}}. \tag{6.1}$$

$\hat{\mathbf{I}}^A$ is the articulated-body inertia and $\hat{\mathbf{p}}$ is the associated bias force, which is the value of the test force which must be applied to the handle in order to give it zero acceleration. It is not always possible to express the relationship between $\hat{\mathbf{f}}$ and $\hat{\mathbf{a}}$ in the form of Equation (6.1), but a sufficient condition is that there be no kinematic connection with ground, i.e. that the articulated body is a 'floating' system. The articulated bodies used in the articulated-body method all satisfy this condition.

The equation of motion of a single rigid body can also be written in the form of Equation (6.1), in which case $\hat{\mathbf{I}}^A$ is its rigid-body inertia and $\hat{\mathbf{p}}$ takes the place of the velocity-product term. It is therefore possible to replace a member of a system of rigid bodies with the handle of an articulated body without affecting the equations of motion of the system. This works provided that the handle is the only point of interaction between the system and the articulated body. However, the real usefulness of articulated-body inertias lies in their power to 'simplify' systems of rigid bodies by allowing several members to be treated as one articulated body, so reducing the apparent number of members in the system.

The following sections outline the articulated-body method, show how to calculate articulated-body inertias, and then describe the articulated-body method in detail.

## 6.2. The Articulated-Body Method in Outline

The basic idea of the articulated-body algorithm is to treat an $n$-joint robot as a one-joint robot whose only moving link is in fact the handle of an articulated body comprising all the remaining links of the robot. To find the acceleration of the first joint requires the solution of the forward dynamics of a one-joint robot, which is much easier than the general case. Having solved for the acceleration of joint 1, we can treat link 1 as the (moving) base of an $n-1$ joint robot and repeat the process for joint 2, and so on.

So to carry out the articulated-body method we need two capabilities: the ability to calculate articulated-body inertias, and the ability to solve the dynamics of a one-joint robot with a moving base. The calculation of articulated-body inertias is described in the next section, so here we shall concentrate on the forward dynamics problem. We could use one of the methods of Chapter 5 to solve this problem, but the one-joint case is so simple that it is easily solved directly.

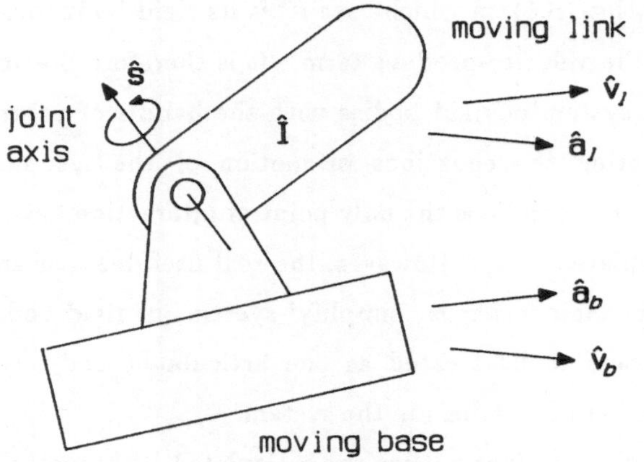


**Figure 6-1:** One-joint robot with moving base

Let a one-joint robot have a base moving with velocity $\hat{\mathbf{v}}_b$ and acceleration $\hat{\mathbf{a}}_b$, a link with inertia $\hat{\mathbf{I}}$, a joint with axis $\hat{\mathbf{s}}$, and let the joint velocity, acceleration and force be $\dot{q}$, $\ddot{q}$ and $Q$ respectively (see Figure 6-1). All quantities are measured in absolute coordinates. The velocity and acceleration of the link are given by

$$\hat{\mathbf{v}}_l = \hat{\mathbf{v}}_b + \hat{\mathbf{s}}\,\dot{q} \tag{6.2}$$

and

$$\hat{\mathbf{a}}_l = \hat{\mathbf{a}}_b + \hat{\mathbf{v}}_b \times \hat{\mathbf{s}}\,\dot{q} + \hat{\mathbf{s}}\,\ddot{q}\ , \tag{6.3}$$

and the equation of motion of the link is

$$\hat{\mathbf{f}} = \hat{\mathbf{I}}\,\hat{\mathbf{a}}_l + \hat{\mathbf{v}}_l \times \hat{\mathbf{I}}\,\hat{\mathbf{v}}_l\,, \tag{6.4}$$

where $\hat{\mathbf{f}}$ is the net force applied to the link (ignoring gravity). $\hat{\mathbf{f}}$ is applied via the joint, so its component on the joint axis is $Q$, i.e.

$$\hat{\mathbf{s}}^S \hat{\mathbf{f}} = Q \,. \tag{6.5}$$

Substituting Equations (6.3 and 6.4) into Equation (6.5), we obtain an expression for $\ddot{q}$ in terms of $Q$:

$$Q = \hat{\mathbf{s}}^S ( \hat{\mathbf{I}} \, \hat{\mathbf{a}}_l + \hat{\mathbf{v}}_l \hat{\times} \hat{\mathbf{I}} \, \hat{\mathbf{v}}_l )$$

$$= \hat{\mathbf{s}}^S ( \hat{\mathbf{I}} ( \hat{\mathbf{a}}_b + \hat{\mathbf{v}}_b \hat{\times} \hat{\mathbf{s}} \dot{q} + \hat{\mathbf{s}} \ddot{q} ) + \hat{\mathbf{v}}_l \hat{\times} \hat{\mathbf{I}} \, \hat{\mathbf{v}}_l ) \,,$$

so

$$\ddot{q} = \frac{Q - \hat{\mathbf{s}}^S ( \hat{\mathbf{I}} ( \hat{\mathbf{a}}_b + \hat{\mathbf{v}}_b \hat{\times} \hat{\mathbf{s}} \dot{q} ) + \hat{\mathbf{v}}_l \hat{\times} \hat{\mathbf{I}} \, \hat{\mathbf{v}}_l )}{\hat{\mathbf{s}}^S \hat{\mathbf{I}} \, \hat{\mathbf{s}}} \,. \tag{6.6}$$

Equation (6.6) is the equation of motion for a one-joint robot with a moving base.

If we replace the single rigid body of the link by an articulated body comprising links 1 . . $n$ of an $n$-joint robot, where link 1 is the handle, and we are able to calculate the articulated-body inertia $\hat{\mathbf{I}}^A$ and the bias force $\hat{\mathbf{p}}$, then we can replace Equation (6.4) with

$$\hat{\mathbf{f}} = \hat{\mathbf{I}}^A \hat{\mathbf{a}}_l + \hat{\mathbf{p}} \tag{6.7}$$

and the equation of motion becomes

$$\ddot{q} = \frac{Q - \hat{\mathbf{s}}^S ( \hat{\mathbf{I}}^A ( \hat{\mathbf{a}}_b + \hat{\mathbf{v}}_b \hat{\times} \hat{\mathbf{s}} \dot{q} ) + \hat{\mathbf{p}})}{\hat{\mathbf{s}}^S \hat{\mathbf{I}}^A \hat{\mathbf{s}}} \,. \tag{6.8}$$

Articulated-body inertias are positive definite, so the denominator is always greater than zero.

Equation (6.8) is the equation of motion of the first joint of an $n$-joint robot. To solve for the motion of the rest of the robot, treat link 1 as the moving base of an $n-1$ joint robot, since we now know its velocity and acceleration, and repeat the whole process. $n$ iterations will solve the problem completely.

## 6.3. The Calculation of Articulated-Body Inertias

Let us start by considering a simple articulated body comprising two members. Calculation methods for more complicated articulated bodies may be obtained by straightforward generalisations of the solution for this particular example.

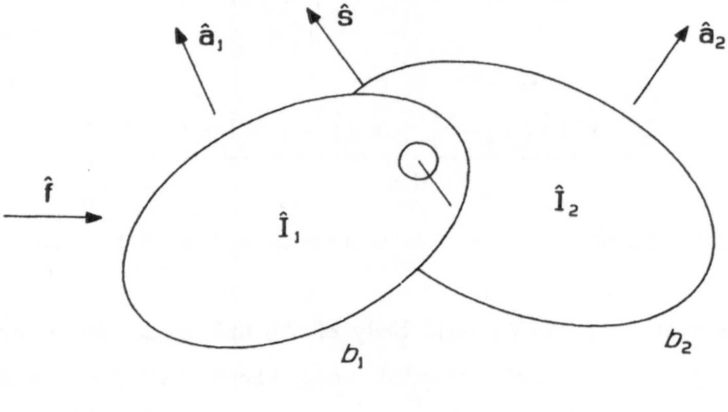


**Figure 6-2:** Simple articulated body

Let the articulated body comprise two members, $b_1$ and $b_2$, having inertias of $\hat{\mathbf{I}}_1$ and $\hat{\mathbf{I}}_2$ respectively and connected by a joint with axis $\hat{\mathbf{s}}$ (see Figure 6-2). The bodies are initially at rest and experience no applied force other than the test force, so the bias force will be zero in this case. A test force $\hat{\mathbf{f}}$ is applied to $b_1$, which is chosen as the handle, causing accelerations of $\hat{\mathbf{a}}_1$ and $\hat{\mathbf{a}}_2$ in $b_1$ and $b_2$ respectively.

Let $\hat{\mathbf{I}}_1^A$ be the articulated-body inertia of $b_1$ in the articulated body comprising $b_1$ and $b_2$; then

$$\hat{\mathbf{f}} = \hat{\mathbf{I}}_1^A \hat{\mathbf{a}}_1 . \tag{6.9}$$

$\hat{\mathbf{f}}$ can be split into two components, $\hat{\mathbf{f}}_1$ and $\hat{\mathbf{f}}_2$, where $\hat{\mathbf{f}}_1$ is the net force acting on $b_1$ and $\hat{\mathbf{f}}_2$ is the force transmitted to $b_2$ through the joint bearings. The equations of motion for the individual bodies are then

$$\hat{\mathbf{f}}_1 = \hat{\mathbf{I}}_1 \hat{\mathbf{a}}_1 \tag{6.10}$$

and

$$\hat{\mathbf{f}}_2 = \hat{\mathbf{I}}_2 \hat{\mathbf{a}}_2 . \tag{6.11}$$

We also have that

$$\hat{\mathbf{f}} = \hat{\mathbf{f}}_1 + \hat{\mathbf{f}}_2 . \tag{6.12}$$

Now the joint imposes constraints on both $\hat{\mathbf{a}}_2$ and $\hat{\mathbf{f}}_2$ which will allow us to express them in terms of $\hat{\mathbf{a}}_1$, and hence allow us to find $\hat{\mathbf{f}}$ in terms of $\hat{\mathbf{a}}_1$. $\hat{\mathbf{a}}_2$ is given by

$$\hat{\mathbf{a}}_2 = \hat{\mathbf{a}}_1 + \hat{\mathbf{s}}\,\alpha , \tag{6.13}$$

where $\alpha$ is the scalar joint acceleration, and $\hat{\mathbf{f}}_2$ does no work in the direction of motion of the joint, so

$$\hat{\mathbf{s}}^S \hat{\mathbf{f}}_2 = 0 . \tag{6.14}$$

We can find $\alpha$ in terms of $\hat{\mathbf{a}}_1$ by substituting Equations (6.11 and 6.13) into Equation (6.14), giving

$$\hat{\mathbf{s}}^S \hat{\mathbf{I}}_2 ( \hat{\mathbf{a}}_1 + \hat{\mathbf{s}}\,\alpha ) = 0$$

$$\alpha = - \frac{\hat{\mathbf{s}}^S \hat{\mathbf{I}}_2 \hat{\mathbf{a}}_1}{\hat{\mathbf{s}}^S \hat{\mathbf{I}}_2 \hat{\mathbf{s}}} . \tag{6.15}$$

Once we know $\alpha$, we can find $\hat{\mathbf{f}}$ in terms of $\hat{\mathbf{a}}_1$:

$$\hat{\mathbf{f}} = \hat{\mathbf{I}}_1 \hat{\mathbf{a}}_1 + \hat{\mathbf{I}}_2 \hat{\mathbf{a}}_2$$

$$= \hat{\mathbf{I}}_1 \hat{\mathbf{a}}_1 + \hat{\mathbf{I}}_2 ( \hat{\mathbf{a}}_1 - \hat{\mathbf{s}} \frac{\hat{\mathbf{s}}^S \hat{\mathbf{I}}_2 \hat{\mathbf{a}}_1}{\hat{\mathbf{s}}^S \hat{\mathbf{I}}_2 \hat{\mathbf{s}}} )$$

$$= (\hat{\mathbf{I}}_1 + \hat{\mathbf{I}}_2 - \frac{\hat{\mathbf{I}}_2 \hat{\mathbf{s}} \hat{\mathbf{s}}^S \hat{\mathbf{I}}_2}{\hat{\mathbf{s}}^S \hat{\mathbf{I}}_2 \hat{\mathbf{s}}}) \hat{\mathbf{a}}_1 . \tag{6.16}$$

Equation (6.16) has the same form as Equation (6.9), and since both are valid for all $\hat{\mathbf{f}}$ it follows that

$$\hat{\mathbf{I}}_1^A = \hat{\mathbf{I}}_1 + \hat{\mathbf{I}}_2 - \frac{\hat{\mathbf{I}}_2 \hat{\mathbf{s}} \hat{\mathbf{s}}^S \hat{\mathbf{I}}_2}{\hat{\mathbf{s}}^S \hat{\mathbf{I}}_2 \hat{\mathbf{s}}} . \tag{6.17}$$

Note two interesting points about Equation (6.17). Firstly, it is a symmetry preserving equation, that is, if $\hat{\mathbf{I}}_1$ and $\hat{\mathbf{I}}_2$ are symmetric (which they are) then $\hat{\mathbf{I}}_1^A$ is also symmetric. Secondly, we can use it to show that $\hat{\mathbf{I}}_1^A$ is bounded below by $\hat{\mathbf{I}}_1$ and bounded above by $\hat{\mathbf{I}}_1 + \hat{\mathbf{I}}_2$ in the sense that for any vector $\hat{\mathbf{x}}$ we have

$$\hat{\mathbf{x}}^S \hat{\mathbf{I}}_1 \hat{\mathbf{x}} \;\le\; \hat{\mathbf{x}}^S \hat{\mathbf{I}}_1^A \hat{\mathbf{x}} \;\le\; \hat{\mathbf{x}}^S (\hat{\mathbf{I}}_1 + \hat{\mathbf{I}}_2) \hat{\mathbf{x}} .$$

This is a stronger statement than merely saying that $\hat{\mathbf{I}}_1^A$ is positive definite, and is proved by letting $\hat{\mathbf{x}} = \hat{\mathbf{x}}_1 + \hat{\mathbf{x}}_2$ where $\hat{\mathbf{x}}_1 = x_1 \hat{\mathbf{s}}$ and $\hat{\mathbf{x}}_2{}^S \hat{\mathbf{I}}_2 \hat{\mathbf{s}} = 0$ (see Chapter 3).

### 6.3.1. Allowing for Velocities and Active Joint Forces

Now let us consider what happens when we allow the bodies to have non-zero velocities and allow an active force to be present at the joint. In this case the bias force will be non-zero, so we must replace Equation (6.9) by the inhomogeneous version

$$\hat{\mathbf{f}} = \hat{\mathbf{I}}_1^A \hat{\mathbf{a}}_1 + \hat{\mathbf{p}}_1 . \tag{6.18}$$

Let $b_1$ and $b_2$ have velocities $\hat{\mathbf{v}}_1$ and $\hat{\mathbf{v}}_2$ respectively (consistent with the joint constraints), then velocity-product terms appear in the equations of motion of $b_1$ and $b_2$, which become

$$\hat{\mathbf{f}}_1 = \hat{\mathbf{I}}_1 \hat{\mathbf{a}}_1 + \hat{\mathbf{p}}_1^v , \tag{6.19}$$

$$\hat{\mathbf{f}}_2 = \hat{\mathbf{I}}_2 \hat{\mathbf{a}}_2 + \hat{\mathbf{p}}_2^v , \tag{6.20}$$

where $\hat{\mathbf{p}}_1^v$ and $\hat{\mathbf{p}}_2^v$ are the bias forces due to velocity-product effects and are given by

$$\hat{\mathbf{p}}_1^v = \hat{\mathbf{v}}_1 \hat{\times} \hat{\mathbf{I}}_1 \hat{\mathbf{v}}_1 , \tag{6.21}$$

$$\hat{\mathbf{p}}_2^v = \hat{\mathbf{v}}_2 \hat{\times} \hat{\mathbf{I}}_2 \hat{\mathbf{v}}_2 . \tag{6.22}$$

A velocity-product term also appears in Equation (6.13) which becomes

$$\hat{\mathbf{a}}_2 = \hat{\mathbf{a}}_1 + \hat{\mathbf{v}}_1 \hat{\times} \hat{\mathbf{v}}_2 + \hat{\mathbf{s}}\,\alpha . \tag{6.23}$$

If we let $Q$ be the active joint force then this appears in Equation (6.14), which becomes

$$\hat{\mathbf{s}}^S \hat{\mathbf{f}}_2 = Q . \tag{6.24}$$

The equations are now a bit more complicated, but we follow exactly the same line of reasoning as before, ending up with the following equations:

$$\hat{\mathbf{I}}_1^A = \hat{\mathbf{I}}_1 + \hat{\mathbf{I}}_2 - \frac{\hat{\mathbf{I}}_2 \hat{\mathbf{s}} \hat{\mathbf{s}}^S \hat{\mathbf{I}}_2}{\hat{\mathbf{s}}^S \hat{\mathbf{I}}_2 \hat{\mathbf{s}}} , \tag{6.25}$$

$$\hat{\mathbf{p}}_1 = \hat{\mathbf{p}}_1^v + \hat{\mathbf{p}}_2^v + \hat{\mathbf{I}}_2 \left( \hat{\mathbf{v}}_1 \hat{\times} \hat{\mathbf{v}}_2 + \hat{\mathbf{s}} \frac{Q - \hat{\mathbf{s}}^S(\hat{\mathbf{I}}_2 \hat{\mathbf{v}}_1 \hat{\times} \hat{\mathbf{v}}_2 + \hat{\mathbf{p}}_2^v)}{\hat{\mathbf{s}}^S \hat{\mathbf{I}}_2 \hat{\mathbf{s}}} \right) . \tag{6.26}$$

Equation (6.25) is the same as Equation (6.17), demonstrating that articulated-body inertias do not depend on the forces present in the system. Equation (6.26) looks complicated, but its evaluation need not be as time-consuming as it might seem since there is some duplication of expressions both within Equation (6.26) and between Equations (6.26 and 6.25).

### 6.3.2. More General Articulated Bodies

The inertia of a more general articulated body can be obtained by replacing the rigid bodies $b_1$ and $b_2$ in the example with the handles of two (independent) articulated bodies. Equations (6.25 and 6.26) then become the general formulae for calculating the inertia of a larger articulated body recursively in terms of the inertias of two smaller articulated bodies. An articulated-body inertia for an articulated body containing $n$ joints can be calculated in $n$ iterations of Equations (6.25 and 6.26), one per joint.

This process is limited to the calculation of inertias for articulated bodies free from kinematic loops. The presence of kinematic loops complicates matters considerably, and consideration of this problem is deferred until Chapter 9.

## 6.4. The Articulated-Body Dynamics Algorithm

We come now to the detailed description of the articulated-body method for forward dynamics, using the system model defined in Chapter 4. The algorithm consists of two steps:

1. the calculation of a series of articulated-body inertias, one for each link, and

2. using the articulated-body inertias to calculate the joint accelerations.

### Step 1: Calculating the articulated-body inertias

Define articulated body $i$ as the articulated body comprising links $i$ . . $n$ of the robot, connected by joints $i+1$ . . $n$ (see Figure 6-3). Link $i$ is always the handle of articulated body $i$. Let $\hat{\mathbf{I}}_i^A$ and $\hat{\mathbf{p}}_i$ be the articulated-body inertia and associated bias force of link $i$ in articulated body $i$. Our target is to find a pair of recurrence relations which will enable us to calculate $\hat{\mathbf{I}}_i^A$ and $\hat{\mathbf{p}}_i$ from $\hat{\mathbf{I}}_{i+1}^A$ and $\hat{\mathbf{p}}_{i+1}$.

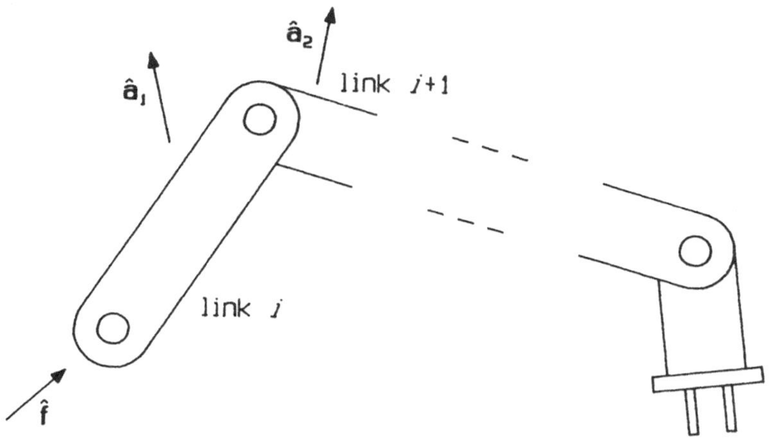


**Figure 6-3:** Articulated body $i$

Since velocity effects are to be included in $\hat{\mathbf{p}}_i$, we must start by calculating the velocities and velocity-product forces at each link in turn. Let $\hat{\mathbf{v}}_i$ be the absolute velocity of link $i$ and let $\hat{\mathbf{p}}_i^v$ be the bias force on link $i$ due to velocity-product forces, then

$$\hat{\mathbf{v}}_i = \hat{\mathbf{v}}_{i-1} + \hat{\mathbf{s}}_i \dot{q}_i \qquad (\hat{\mathbf{v}}_0 = \hat{\mathbf{0}}) \tag{6.27}$$

and

$$\hat{\mathbf{p}}_i^v = \hat{\mathbf{v}}_i \hat{\times} \hat{\mathbf{I}}_i \hat{\mathbf{v}}_i \,. \tag{6.28}$$

Terms due to external forces acting on link $i$ may also be included in Equation (6.28). The equation of motion for link $i$ then takes the form

$$\hat{\mathbf{f}} = \hat{\mathbf{I}}_i \hat{\mathbf{a}} + \hat{\mathbf{p}}_i^v \,.$$

Apply a test force $\hat{\mathbf{f}}$ to link $i$ in articulated body $i$, producing an acceleration of $\hat{\mathbf{a}}_1$ in link $i$ and $\hat{\mathbf{a}}_2$ in link $i+1$. Splitting $\hat{\mathbf{f}}$ into components $\hat{\mathbf{f}}_1$ and $\hat{\mathbf{f}}_2$, where $\hat{\mathbf{f}}_1$ acts on link $i$ and $\hat{\mathbf{f}}_2$ is passed on to link $i+1$, and treating link $i+1$ as the handle of articulated body $i+1$, we have the following relations:

$$\hat{\mathbf{f}} = \hat{\mathbf{I}}_i^A \, \hat{\mathbf{a}}_1 + \hat{\mathbf{p}}_i \,, \tag{6.29}$$

$$\hat{\mathbf{f}}_1 = \hat{\mathbf{I}}_i \, \hat{\mathbf{a}}_1 + \hat{\mathbf{p}}_i^v \,, \tag{6.30}$$

$$\hat{\mathbf{f}}_2 = \hat{\mathbf{I}}_{i+1}^A \, \hat{\mathbf{a}}_2 + \hat{\mathbf{p}}_{i+1} \tag{6.31}$$

and

$$\hat{\mathbf{f}} = \hat{\mathbf{f}}_1 + \hat{\mathbf{f}}_2 \,. \tag{6.32}$$

Joint $i$+1 constrains $\hat{\mathbf{a}}_2$ such that it can be described in terms of $\hat{\mathbf{a}}_1$ and a single unknown scalar, and it imposes one degree of constraint on $\hat{\mathbf{f}}_2$. $\hat{\mathbf{a}}_2$ is given by

$$\hat{\mathbf{a}}_2 = \hat{\mathbf{a}}_1 + \hat{\mathbf{v}}_{i+1} \hat{\times} \hat{\mathbf{s}}_{i+1} \dot{q}_{i+1} + \hat{\mathbf{s}}_{i+1} \, \alpha \,, \tag{6.33}$$

where $\alpha$ is the unknown joint acceleration; and the component of $\hat{\mathbf{f}}_2$ on the axis of joint $i$+1 is $Q_{i+1}$; i.e.,

$$\hat{\mathbf{s}}_{i+1}^S \, \hat{\mathbf{f}}_2 = Q_{i+1} \,. \tag{6.34}$$

We need to find $\alpha$ in terms of $\hat{\mathbf{a}}_1$. This can be done by substituting Equations (6.31 and 6.33) into Equation (6.34), giving

$$\hat{\mathbf{s}}_{i+1}^S (\hat{\mathbf{I}}_{i+1}^A (\hat{\mathbf{a}}_1 + \hat{\mathbf{v}}_{i+1} \hat{\times} \hat{\mathbf{s}}_{i+1} \dot{q}_{i+1} + \hat{\mathbf{s}}_{i+1} \, \alpha) + \hat{\mathbf{p}}_{i+1}) = Q_{i+1} \,,$$

from which it follows that

$$\alpha = \frac{Q_{i+1} - \hat{\mathbf{s}}_{i+1}^S (\hat{\mathbf{I}}_{i+1}^A (\hat{\mathbf{a}}_1 + \hat{\mathbf{v}}_{i+1} \hat{\times} \hat{\mathbf{s}}_{i+1} \dot{q}_{i+1}) + \hat{\mathbf{p}}_{i+1})}{\hat{\mathbf{s}}_{i+1}^S \, \hat{\mathbf{I}}_{i+1}^A \, \hat{\mathbf{s}}_{i+1}} \,. \tag{6.35}$$

Now we know $\alpha$, we can express $\hat{\mathbf{a}}_2$ in terms of $\hat{\mathbf{a}}_1$ and hence we can express $\hat{\mathbf{f}}$ in terms of $\hat{\mathbf{a}}_1$ using Equations (6.30-6.33).

$$\hat{\mathbf{f}} = \hat{\mathbf{I}}_i \hat{\mathbf{a}}_1 + \hat{\mathbf{p}}_i^v + \hat{\mathbf{I}}_{i+1}^A \hat{\mathbf{a}}_2 + \hat{\mathbf{p}}_{i+1}$$

$$= \hat{\mathbf{I}}_i \hat{\mathbf{a}}_1 + \hat{\mathbf{p}}_i^v + \hat{\mathbf{I}}_{i+1}^A (\hat{\mathbf{a}}_1 + \hat{\mathbf{v}}_{i+1} \hat{\times} \hat{\mathbf{s}}_{i+1} \dot{q}_{i+1} + \hat{\mathbf{s}}_{i+1} \alpha) + \hat{\mathbf{p}}_{i+1}$$

$$= \hat{\mathbf{I}}_i \hat{\mathbf{a}}_1 + \hat{\mathbf{p}}_i^v + \hat{\mathbf{p}}_{i+1} + \hat{\mathbf{I}}_{i+1}^A (\hat{\mathbf{a}}_1 + \hat{\mathbf{v}}_{i+1} \hat{\times} \hat{\mathbf{s}}_{i+1} \dot{q}_{i+1}$$

$$+ \hat{\mathbf{s}}_{i+1} \frac{Q_{i+1} - \hat{\mathbf{s}}_{i+1}^S (\hat{\mathbf{I}}_{i+1}^A (\hat{\mathbf{a}}_1 + \hat{\mathbf{v}}_{i+1} \hat{\times} \hat{\mathbf{s}}_{i+1} \dot{q}_{i+1}) + \hat{\mathbf{p}}_{i+1})}{\hat{\mathbf{s}}_{i+1}^S \hat{\mathbf{I}}_{i+1}^A \hat{\mathbf{s}}_{i+1}})$$

$$= (\hat{\mathbf{I}}_i + \hat{\mathbf{I}}_{i+1}^A - \frac{\hat{\mathbf{I}}_{i+1}^A \hat{\mathbf{s}}_{i+1} \hat{\mathbf{s}}_{i+1}^S \hat{\mathbf{I}}_{i+1}^A}{\hat{\mathbf{s}}_{i+1}^S \hat{\mathbf{I}}_{i+1}^A \hat{\mathbf{s}}_{i+1}}) \hat{\mathbf{a}}_1$$

$$+ \hat{\mathbf{p}}_i^v + \hat{\mathbf{p}}_{i+1} + \hat{\mathbf{I}}_{i+1}^A \hat{\mathbf{v}}_{i+1} \hat{\times} \hat{\mathbf{s}}_{i+1} \dot{q}_{i+1}$$

$$+ \frac{\hat{\mathbf{I}}_{i+1}^A \hat{\mathbf{s}}_{i+1} (Q_{i+1} - \hat{\mathbf{s}}_{i+1}^S (\hat{\mathbf{I}}_{i+1}^A \hat{\mathbf{v}}_{i+1} \hat{\times} \hat{\mathbf{s}}_{i+1} \dot{q}_{i+1} + \hat{\mathbf{p}}_{i+1}))}{\hat{\mathbf{s}}_{i+1}^S \hat{\mathbf{I}}_{i+1}^A \hat{\mathbf{s}}_{i+1}}. \quad (6.36)$$

On comparing Equation (6.36) with Equation (6.29) we get the following expressions for $\hat{\mathbf{I}}_i^A$ and $\hat{\mathbf{p}}_i$:

$$\hat{\mathbf{I}}_i^A = \hat{\mathbf{I}}_i + \hat{\mathbf{I}}_{i+1}^A - \frac{\hat{\mathbf{I}}_{i+1}^A \hat{\mathbf{s}}_{i+1} \hat{\mathbf{s}}_{i+1}^S \hat{\mathbf{I}}_{i+1}^A}{\hat{\mathbf{s}}_{i+1}^S \hat{\mathbf{I}}_{i+1}^A \hat{\mathbf{s}}_{i+1}}, \quad (6.37)$$

$$\hat{\mathbf{p}}_i = \hat{\mathbf{p}}_i^v + \hat{\mathbf{p}}_{i+1} + \hat{\mathbf{I}}_{i+1}^A \hat{\mathbf{v}}_{i+1} \hat{\times} \hat{\mathbf{s}}_{i+1} \dot{q}_{i+1}$$

$$+ \frac{\hat{\mathbf{I}}_{i+1}^A \hat{\mathbf{s}}_{i+1} (Q_{i+1} - \hat{\mathbf{s}}_{i+1}^S (\hat{\mathbf{I}}_{i+1}^A \hat{\mathbf{v}}_{i+1} \hat{\times} \hat{\mathbf{s}}_{i+1} \dot{q}_{i+1} + \hat{\mathbf{p}}_{i+1}))}{\hat{\mathbf{s}}_{i+1}^S \hat{\mathbf{I}}_{i+1}^A \hat{\mathbf{s}}_{i+1}}. \quad (6.38)$$

The starting values for these recurrence relations are simply

$$\hat{\mathbf{I}}_n^A = \hat{\mathbf{I}}_n \quad (6.39)$$

and

$$\hat{\mathbf{p}}_n = \hat{\mathbf{p}}_n^v, \tag{6.40}$$

since articulated body $n$ is just a single rigid body.

**Step 2: Calculating the joint accelerations**

Having found all the articulated-body inertias, we can now start calculating the joint accelerations. At the $i^{th}$ stage of this computation, link $i-1$ is treated as the moving base of a one-joint robot and articulated body $i$ as its only apparent link. So, if we let $\hat{\mathbf{a}}_i$ be the absolute acceleration of link $i$, we calculate $\ddot{q}_i$ assuming that $\hat{\mathbf{a}}_{i-1}$ is known, which then allows us to calculate $\hat{\mathbf{a}}_i$ ready for the next iteration.

$\hat{\mathbf{a}}_i$ is given in terms of $\hat{\mathbf{a}}_{i-1}$ by

$$\hat{\mathbf{a}}_i = \hat{\mathbf{a}}_{i-1} + \hat{\mathbf{v}}_i \hat{\times} \hat{\mathbf{s}}_i \dot{q}_i + \hat{\mathbf{s}}_i \ddot{q}_i, \tag{6.41}$$

and the total force, $\hat{\mathbf{f}}$, transmitted through joint $i$ is related to $\hat{\mathbf{a}}_i$ by

$$\hat{\mathbf{f}} = \hat{\mathbf{I}}_i^A \hat{\mathbf{a}}_i + \hat{\mathbf{p}}_i$$

$$= \hat{\mathbf{I}}_i^A (\hat{\mathbf{a}}_{i-1} + \hat{\mathbf{v}}_i \hat{\times} \hat{\mathbf{s}}_i \dot{q}_i + \hat{\mathbf{s}}_i \ddot{q}_i) + \hat{\mathbf{p}}_i. \tag{6.42}$$

The component of $\hat{\mathbf{f}}$ on the axis of joint $i$ is $Q_i$, so we have

$$Q_i = \hat{\mathbf{s}}_i^S \hat{\mathbf{f}}$$

$$= \hat{\mathbf{s}}_i^S (\hat{\mathbf{I}}_i^A (\hat{\mathbf{a}}_{i-1} + \hat{\mathbf{v}}_i \hat{\times} \hat{\mathbf{s}}_i \dot{q}_i + \hat{\mathbf{s}}_i \ddot{q}_i) + \hat{\mathbf{p}}_i),$$

giving

$$\ddot{q}_i = \frac{Q_i - \hat{\mathbf{s}}_i^S (\hat{\mathbf{I}}_i^A (\hat{\mathbf{a}}_{i-1} + \hat{\mathbf{v}}_i \hat{\times} \hat{\mathbf{s}}_i \dot{q}_i) + \hat{\mathbf{p}}_i)}{\hat{\mathbf{s}}_i^S \hat{\mathbf{I}}_i^A \hat{\mathbf{s}}_i}. \tag{6.43}$$

Substituting the value for $\ddot{q}_i$ back into Equation (6.41) gives us $\hat{\mathbf{a}}_i$ for the next iteration, and the starting value for Equation (6.41) is

$$\hat{\mathbf{a}}_0 = \hat{\mathbf{0}}. \tag{6.44}$$

Gravity may be catered for by giving the base a fictitious acceleration of $-\hat{\mathbf{g}}$, where $\hat{\mathbf{g}}$ is the gravitational acceleration vector, in which case the starting value for Equation (6.41) is

$$\hat{\mathbf{a}}_0 = -\hat{\mathbf{g}}. \tag{6.45}$$

### 6.4.1. Common Sub-Expressions

If the articulated-body algorithm is carried out exactly in accordance with the above equations then a significant amount of duplicate calculation is performed due to the presence of common sub-expressions both within and between the equations. This repetition may be eliminated by introducing the following partial results which should be computed once and stored for later use:

$$\hat{\mathbf{c}}_i = \hat{\mathbf{v}}_i \hat{\times} \hat{\mathbf{s}}_i \dot{q}_i\,, \tag{6.46}$$

$$\hat{\mathbf{h}}_i = \hat{\mathbf{I}}_i^A \hat{\mathbf{s}}_i\,, \tag{6.47}$$

$$d_i = \hat{\mathbf{s}}_i^S \hat{\mathbf{h}}_i \tag{6.48}$$

and

$$u_i = Q_i - \hat{\mathbf{h}}_i^S \hat{\mathbf{c}}_i - \hat{\mathbf{s}}_i^S \hat{\mathbf{p}}_i\,. \tag{6.49}$$

In terms of these partial results, the equations of the articulated-body algorithm become:

$$\hat{\mathbf{v}}_i = \hat{\mathbf{v}}_{i-1} + \hat{\mathbf{s}}_i \dot{q}_i\,, \qquad (\hat{\mathbf{v}}_0 = \hat{\mathbf{0}}) \tag{6.50}$$

$$\hat{\mathbf{p}}_i^v = \hat{\mathbf{v}}_i \hat{\times} \hat{\mathbf{I}}_i \hat{\mathbf{v}}_i\,, \tag{6.51}$$

$$\hat{\mathbf{I}}_i^A = \hat{\mathbf{I}}_i + \hat{\mathbf{I}}_{i+1}^A - \frac{\hat{\mathbf{h}}_{i+1}\,\hat{\mathbf{h}}_{i+1}^S}{d_{i+1}}, \qquad (\hat{\mathbf{I}}_n^A = \hat{\mathbf{I}}_n) \tag{6.52}$$

$$\hat{\mathbf{p}}_i = \hat{\mathbf{p}}_i^v + \hat{\mathbf{p}}_{i+1} + \hat{\mathbf{I}}_{i+1}^A\,\hat{\mathbf{c}}_{i+1} + \frac{u_{i+1}}{d_{i+1}}\hat{\mathbf{h}}_{i+1}, \quad (\hat{\mathbf{p}}_n = \hat{\mathbf{p}}_n^v) \tag{6.53}$$

$$\hat{\mathbf{a}}_i = \hat{\mathbf{a}}_{i-1} + \hat{\mathbf{c}}_i + \hat{\mathbf{s}}_i\,\ddot{q}_i, \qquad (\hat{\mathbf{a}}_0 = \hat{\mathbf{0}}) \tag{6.54}$$

$$\ddot{q}_i = \frac{u_i - \hat{\mathbf{h}}_i^S\,\hat{\mathbf{a}}_{i-1}}{d_i}. \tag{6.55}$$

### 6.4.2. Computation in Link Coordinates

There is a slight benefit from performing this computation in link coordinates, although it depends entirely on the greater scope for optimisation in link coordinates than in absolute coordinates (see Chapter 8). The conversion to link coordinates is effected by straightforward application of the rules laid down in Chapter 4, and result in the following equations:

$$\hat{\mathbf{c}}_i = \hat{\mathbf{v}}_i \hat{\times} \hat{\mathbf{s}}_i'\,\dot{q}_i, \tag{6.56}$$

$$\hat{\mathbf{h}}_i = \hat{\mathbf{I}}_i^A\,\hat{\mathbf{s}}_i', \tag{6.57}$$

$$d_i = \hat{\mathbf{s}}_i'^S\,\hat{\mathbf{h}}_i, \tag{6.58}$$

$$u_i = Q_i - \hat{\mathbf{h}}_i^S\,\hat{\mathbf{c}}_i - \hat{\mathbf{s}}_i'^S\,\hat{\mathbf{p}}_i, \tag{6.59}$$

$$\hat{\mathbf{v}}_i = {}_i\hat{\mathbf{X}}_{i-1}\,\hat{\mathbf{v}}_{i-1} + \hat{\mathbf{s}}_i'\,\dot{q}_i, \qquad (\hat{\mathbf{v}}_0 = \hat{\mathbf{0}}) \tag{6.60}$$

$$\hat{\mathbf{p}}_i^v = \hat{\mathbf{v}}_i \hat{\times} \hat{\mathbf{I}}_i' \hat{\mathbf{v}}_i \,, \tag{6.61}$$

$$\hat{\mathbf{I}}_i^A = \hat{\mathbf{I}}_i' + {}_i\hat{\mathbf{X}}_{i+1}(\hat{\mathbf{I}}_{i+1}^A - \frac{\hat{\mathbf{h}}_{i+1}\,\hat{\mathbf{h}}_{i+1}^S}{d_{i+1}})\,{}_{i+1}\hat{\mathbf{X}}_i \,, \quad (\hat{\mathbf{I}}_n^A = \hat{\mathbf{I}}_n') \tag{6.62}$$

$$\hat{\mathbf{p}}_i = \hat{\mathbf{p}}_i^v + {}_i\hat{\mathbf{X}}_{i+1}(\hat{\mathbf{p}}_{i+1} + \hat{\mathbf{I}}_{i+1}^A\,\hat{\mathbf{c}}_{i+1} + \frac{u_{i+1}}{d_{i+1}}\hat{\mathbf{h}}_{i+1}) \,, \quad (\hat{\mathbf{p}}_n = \hat{\mathbf{p}}_n^v) \tag{6.63}$$

$$\hat{\mathbf{a}}_i = {}_i\hat{\mathbf{X}}_{i-1}\,\hat{\mathbf{a}}_{i-1} + \hat{\mathbf{c}}_i + \hat{\mathbf{s}}_i'\,\ddot{q}_i \,, \qquad (\hat{\mathbf{a}}_0 = \hat{\mathbf{0}}) \tag{6.64}$$

$$\ddot{q}_i = \frac{u_i - \hat{\mathbf{h}}_i^S\,{}_i\hat{\mathbf{X}}_{i-1}\,\hat{\mathbf{a}}_{i-1}}{d_i} \,. \tag{6.65}$$

# Chapter 7
# Extending the Dynamics Algorithms

## 7.1. Introduction

In this chapter we will be concerned with generalisations of the basic dynamics algorithms described in Chapters 4 to 6. The system model defined in Chapter 4 will be extended to allow the description of a wider class of robot mechanisms, and the dynamics algorithms will be modified accordingly. The material is organised into three topics:

- generalising the robot's connectivity to allow branched kinematic chains,
- allowing screw joints and joints with more than one degree of freedom, and
- allowing the base member to move so that its acceleration is not known in advance.

For simplicity, each extension is described independently of the others, although there is no difficulty in implementing them in combination. Wherever possible, modifications to the basic dynamics algorithms are described in terms of rules for modifying the equations presented in Chapters 4 to 6 rather than by re-stating or re-deriving entire algorithms.

The implementation of branched kinematic chains is accomplished by including an explicit description of the link connectivity as part of the system model, and making a few systematic modifications to the equations of the dynamics algorithms. The properties and performances of the algorithms are not affected significantly. An example of a branched kinematic chain is a robot with a multi-fingered gripper. Kinematic loops are specifically excluded at this stage since they introduce many more problems than tree structures, so we will not yet be able to cope with a multi-fingered gripper manipulating an object. Kinematic loops will be considered in Chapter 9.

Screw joints can be implemented without any change to the system model or the dynamics algorithms simply by allowing the joint axis vector to represent a screw axis. They do, however, raise the issue of magnitudes of unit screw forces and displacements. This issue is clarified under multiple-degree-of-freedom joints with the introduction of explicit actuator models.

Multiple-degree-of-freedom joints introduce more significant changes to both the system model and the dynamics algorithms: the notion of a powered joint needs to be defined more carefully, and a number of quantities in the dynamics algorithms which were previously scalars become vectors or matrices. Although multiple-degree-of-freedom joints can be synthesised from the appropriate number of single-degree-of-freedom joints, this is not always convenient and can lead to computational inefficiencies.

A moving-base robot can be treated as a fixed-base robot by introducing a fictitious, un-powered joint between the base and some fixed point. If the base has complete motion freedom then the joint has six degrees of freedom. The forward dynamics algorithms cope with this easily enough, but inverse dynamics needs significant modification because of the unknown base acceleration. The modified algorithm is a mixture of forward and inverse dynamics. Examples of robots with moving bases include robots mounted on spacecraft and legged robots leaping through the air.

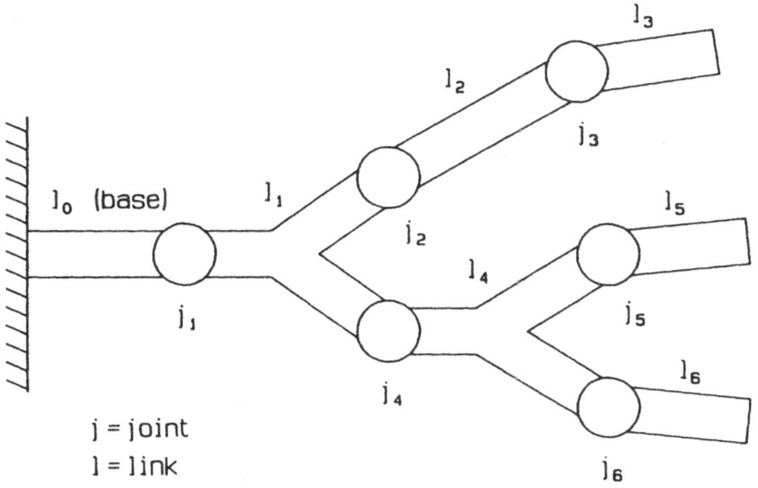


**Figure 7-1:** Robot mechanism with branched chain structure

## 7.2. Branched Kinematic Chains

By a branched kinematic chain I mean that the links may be connected in any manner corresponding to a topological tree. Basically, this means that there are no kinematic loops and that no part of the robot is entirely disconnected from the rest. Figure 7-1 gives an example. The base link (which is still considered immobile) is chosen as the root of the tree, and the outermost links are its leaves. Two links connected by a joint are neighbours, and a neighbour of a given link will be called its predecessor or its successor depending on whether it is nearer to or further from the base. The joint connecting a link to its predecessor will be called its inner joint, and those connecting a link to its successors its outer joints. Every link except the base has exactly one inner joint, but may have any number of outer joints.

The main difference between such a tree structure and the un-branched chains we have been using up to now is that for the tree structure we need to specify the connectivity explicitly whereas for the un-branched chain it was implicit in the numbering scheme. The connectivity is defined by introducing for each link the quantity $\lambda_i$, which is the identification number of the preceding link.[12] For the tree structure shown in Figure 7-1 the values of $\lambda_i$ for $i=1 \ . \ . \ 6$ are (0,1,2,1,4,4). The $\lambda_i$ alone are enough to define the connectivity, but for expressing iterations going from the base to the leaves it is convenient to introduce for each link the additional quantity $\mu_i$, which is the set of identification numbers of all the successors of link $i$. For the tree structure shown in Figure 7-1 the values of $\mu_i$ are $\mu_0=\{1\}$, $\mu_1=\{2,4\}$, $\mu_2=\{3\}$, $\mu_4=\{5,6\}$, $\mu_3,\mu_5,\mu_6=\{\}$. The leaves are distinguished by having empty $\mu$'s.

A systematic enumeration of the links is no longer necessary, but it is still convenient and will be retained. The numbering scheme we shall use is to call the base link 0, and number the other links from 1 to $n$ so that every link has a higher identification number than its predecessor. The inner joint for link $i$ is called joint $i$. (The numbering on Figure 7-1 follows this scheme.) This numbering scheme degenerates into our original numbering scheme when applied to an un-branched kinematic chain, but there is not normally a unique numbering for a branched kinematic chain. The adjacent-link coordinate transformation across joint $i$ is now ${}_{\lambda_i}\hat{\mathbf{X}}_i$.

### 7.2.1. Modifying the Algorithms

The equations of the Newton-Euler and articulated-body algorithms may be classified into three groups: those that iterate from the base to the end effector, those that iterate from the end effector to the base, and those that perform computations local to each link. Members of the last group are independent of the connectivity and need no modification.

---

[12] These are the elements of the body connection array defined by Huston, et al. [29]. An alternative approach, not considered here, is to use an incidence matrix [76].

Outward iterating equations carry motion-type quantities from the base to the end effector. An example is the link velocity calculation of Equation (4.6):

$$\hat{v}_i = \hat{v}_{i-1} + \hat{s}_i \dot{q}_i . \qquad (\hat{v}_0 = \hat{0}) \tag{7.1}$$

In this context the expression $i-1$ really means 'the predecessor of link $i$', so modifying these equations to work on branched kinematic chains is simply a matter of replacing instances of $i-1$ with $\lambda_i$. The link velocity calculation for a tree structure robot becomes

$$\hat{v}_i = \hat{v}_{\lambda_i} + \hat{s}_i \dot{q}_i . \qquad (\hat{v}_0 = \hat{0}) \tag{7.2}$$

If we stick to the original order of calculation, where $i$ goes from 1 to $n$, then $\hat{v}_{\lambda_i}$ will always be calculated before $\hat{v}_i$ (since $\lambda_i < i$).

Inward iterating equations carry forces or inertias from the end effector to the base. With a tree structure, any given link receives one such quantity from each of its successors, and since these quantities are additive the total amount is the sum of all the contributions. For example, the calculation of total joint forces for the Newton-Euler method (Equation (4.9)) is given by

$$\hat{f}_i = \hat{f}_{i+1} + \hat{f}^*_i . \qquad (\hat{f}_n = \hat{f}^*_n) \tag{7.3}$$

This equation states that the total force required to support the motion of all links from $i$ onwards is the sum of the forces required to support the motion of link $i$ and the motion of all following links. In the context of a tree structure, link $i$ supports directly all links $j$ where $j \in \mu_i$, so $\hat{f}_{i+1}$ in Equation (7.3) must be replaced by $\sum_{j \in \mu_i} \hat{f}_j$ giving

$$\hat{f}_i = \sum_{j \in \mu_i} \hat{f}_j + \hat{f}^*_i . \tag{7.4}$$

Notice that there is now no need for a separate initial condition, since the calculation of $\hat{f}_i$ when link $i$ is a leaf is handled correctly by virtue of $\mu_i$

being the empty set for such a link. Again, if we stick to the original order of calculation, where $i$ goes from $n$ down to 1, we will always calculate all the $\hat{\mathbf{f}}_j$ before $\hat{\mathbf{f}}_i$.

A more complicated example is the calculation of articulated-body inertias. For an un-branched chain this is given by Equation (6.37):

$$\hat{\mathbf{I}}_i^A = \hat{\mathbf{I}}_i + \hat{\mathbf{I}}_{i+1}^A - \frac{\hat{\mathbf{I}}_{i+1}^A \, \hat{\mathbf{s}}_{i+1} \, \hat{\mathbf{s}}_{i+1}^S \, \hat{\mathbf{I}}_{i+1}^A}{\hat{\mathbf{s}}_{i+1}^S \, \hat{\mathbf{I}}_{i+1}^A \, \hat{\mathbf{s}}_{i+1}} . \quad ( \hat{\mathbf{I}}_n^A = \hat{\mathbf{I}}_n ) \tag{7.5}$$

The quantity that gets passed back from the successor link appears as an expression rather than a single term, so the branched-chain version of this equation is

$$\hat{\mathbf{I}}_i^A = \hat{\mathbf{I}}_i + \sum_{j \in \mu_i} ( \hat{\mathbf{I}}_j^A - \frac{\hat{\mathbf{I}}_j^A \, \hat{\mathbf{s}}_j \, \hat{\mathbf{s}}_j^S \, \hat{\mathbf{I}}_j^A}{\hat{\mathbf{s}}_j^S \, \hat{\mathbf{I}}_j^A \, \hat{\mathbf{s}}_j} ) . \tag{7.6}$$

The order of traversal of the tree can be controlled by the numbering scheme. This has no effect on the amount of computation performed, but it can affect the amount of temporary storage needed.

The computation of the joint-space inertia matrix does not fit into the scheme outlined above, but it is still easily modified. The joint-space inertia matrix is defined by Equation (5.9) as

$$H_{ji} = \hat{\mathbf{s}}_j^S \hat{\mathbf{f}}_i , \qquad ( j \leq i ) \tag{7.7}$$

where $\hat{\mathbf{f}}_i$ is the force required to impart unit acceleration about joint $i$ to the composite rigid body comprising links $i \, . \, . \, n$, and $H_{ji}$ is the joint force required at joint $j$ to support $\hat{\mathbf{f}}_i$ at joint $i$. In the case of a tree structure, this equation only applies if joint $j$ is supporting the sub-tree containing the moving links. For example, referring back to Figure 7-1, if $i$=5 then Equation (7.7) only applies to joints 1, 4 and 5, and the values of $H_{25}$ and $H_{35}$ are zero since neither joint 2 nor joint 3 supports the motion of joint 5.

To express this concisely, let us introduce the quantities $\nu_i$ which, for each joint, list all the links in the sub-tree supported by that joint. $\nu_i$ is defined by

$$j \in \nu_i \text{ iff } j = i \text{ or } ( j \in \mu_k \text{ and } k \in \nu_i ) . \tag{7.8}$$

For the tree structure shown in Figure 7-1, $\nu_1=\{1,2,3,4,5,6\}$, $\nu_2=\{2,3\}$, and so on. Armed with this definition, we may express the tree structure version of Equation (7.7) as

$$H_{ji} = \begin{cases} \hat{\mathbf{s}}_j^S \hat{\mathbf{f}}_i & \text{if } j \in \nu_i \\ 0 & \text{if } j \notin \nu_i \end{cases} . \tag{7.9}$$

The quantities $\nu_i$ are not necessary for calculating **H**. The simplest calculation scheme is to set every element of **H** initially to zero, and use Equation (7.7) with $i$ ranging from $n$ down to 1 and $j$ taking the values $i$, $\lambda_i$, $\lambda_{\lambda_i}$, etc., to compute the non-zero elements of **H**.

## 7.3. Screw Joints

The representation of screw joint axes was discussed in Chapter 2 along with revolute and prismatic ones. Although the system model defined in Chapter 4 specified that joints should be either revolute or prismatic, they can just as easily be screw joints, without affecting any of the dynamics equations, simply by allowing $\hat{\mathbf{s}}_i'$ to represent a screw axis.

The issue raised by screw joints is really one of units and magnitudes. We have assumed that the unit joint force acting over the unit joint displacement performs the unit amount of work. Moreover, we have assumed that the joint force for a revolute joint is a pure couple in the appropriate direction and that for a prismatic joint a pure force. The most logical choice of joint force for a screw joint of pitch $p$ is a screw force of pitch $1/p$; but if we use the standard definition of a unit screw, which is that the line-vector part should have unit magnitude, then the unit screw force acting over the unit screw displacement performs $\rho+1/\rho$ times the unit of work. So we define the unit joint force for a screw joint of pitch $\rho$ to be a screw force of magnitude $\rho/(1+\rho^2)$ and pitch $1/\rho$. If the unit joint displacement is a rotation of one radian about the joint axis plus a translation of $\rho$ metres along it, then the unit joint force is a force of $\rho/(1+\rho^2)$ Newtons along the joint axis plus a couple of $1/(1+\rho^2)$ Newton-metres about it.

It is tempting to say that the standard definition of a unit screw is inappropriate, and that a better definition is to set the magnitude of the line-vector part of a screw of pitch $\rho$ to $(\rho/(1+\rho^2))^{1/2}$, but this definition is based on the choice of $1/\rho$ as the pitch of the joint force. We could instead choose the joint force for a screw joint to be a pure couple, in which case the unit joint force would always be a unit couple, regardless of the pitch.

The situation will be made clearer in the next section, which introduces an explicit model of the actuation system and relaxes the condition that the unit joint force acting over the unit joint displacement should perform one unit of work.

## 7.4. Multiple-Degree-of-Freedom Joints

A multiple-degree-of-freedom joint can be simulated by a chain of the appropriate number of single-degree-of-freedom joints; but there are occasions when this can be inconvenient or lead to computational inefficiencies, and the introduction of fictitious links with zero inertia can cause numerical problems in the dynamics computations. In order to represent multiple-degree-of-freedom joints directly, the joint model must be modified so that the joint axis is replaced by a motion sub-space and the scalar joint variables by vectors[13] of the appropriate dimension. Also, it is convenient at this stage to introduce an explicit model of the actuation system for a powered joint.

### 7.4.1. The Joint Model

The type of motion allowed by a single-degree-of-freedom joint is defined by its axis, $\hat{\mathbf{s}}$, and the spatial joint velocity is given by $\hat{\mathbf{v}} = \hat{\mathbf{s}}\,\dot{q}$, where $\dot{q}$ is the joint velocity variable. The instantaneous motion allowed by an $r$-degree-of-freedom joint is defined by the $r$-dimensional motion sub-space $\hat{\mathbf{S}}$, which is, in general, a function of the joint displacement.

If joint $j$ is a multiple-degree-of-freedom joint having $n_j$ degrees of freedom, then the generalised velocity of the joint is given by the $n_j$-dimensional vector $\dot{\mathbf{q}}_j$, and the spatial velocity of the joint by

$$\hat{\mathbf{v}} = \hat{\mathbf{S}}_j \dot{\mathbf{q}}_j . \tag{7.10}$$

The columns of $\hat{\mathbf{S}}_j$ are important, since they define the motions to which each element of $\dot{\mathbf{q}}_j$ refers. The generalised acceleration of the joint is given by the $n_j$-dimensional vector $\ddot{\mathbf{q}}_j$ and the spatial acceleration by

$$\hat{\mathbf{a}} = \hat{\mathbf{S}}_j \ddot{\mathbf{q}}_j + \dot{\hat{\mathbf{S}}}_j \dot{\mathbf{q}}_j . \tag{7.11}$$

For many common kinds of joint it is possible to arrange for $\hat{\mathbf{S}}_j$ to be constant, in which case the term involving $\dot{\hat{\mathbf{S}}}_j$ (which is effectively an additional velocity-product term) does not appear.

The joint displacement is given by the $n_j$-dimensional array of joint variables $\mathbf{q}_j$. The adjacent-link transformation across joint $j$ is a function of $\mathbf{q}_j$. The details of how to compute the adjacent-link transformation from $\mathbf{q}_j$ depend on the type of motion allowed by the joint, and will not be

[13] Strictly speaking, the joint position variables do not constitute the elements of a vector.

considered here. Some examples are described in Chapter 8, along with examples of joint motion sub-spaces.

With the introduction of multiple-degree-of-freedom joints, the number of degrees of freedom of the system as a whole will be different from the number of joints. If $n$ is the number of degrees of freedom of the system, $n_j$ the number of degrees of freedom for joint $j$, and $N$ the number of joints, then

$$n = \sum_{j=1}^{N} n_j .$$

The vector of system generalised coordinates, $\mathbf{q}$, is now an $n$-dimensional vector formed by concatenating the vectors of joint variables. Strictly speaking, $\mathbf{q}_j$ is a subvector of $\mathbf{q}$; but for simplicity I shall continue to refer to $\mathbf{q}_j$ as the $j^{th}$ component of $\mathbf{q}$, and similarly for generalised velocities, etc..

### 7.4.2. The Actuator Model

A powered joint is deemed to have an actuation system which is capable of generating spatial forces in the sub-space $\hat{\mathbf{F}}$. As with $\hat{\mathbf{S}}$, $\hat{\mathbf{F}}$ is in general a function of the joint variables, but it can be arranged to be constant for most common kinds of joint. For a joint to be 'properly powered' the actuation space must be of the same dimension as the motion space, and it must not have any member in common with the reaction-force space $\hat{\mathbf{S}}^{\perp}$. In other words, $\hat{\mathbf{S}}^S \hat{\mathbf{F}}$ must be non-singular. It is possible to conceive of joints where the actuation system has fewer degrees of freedom than the joint (under powered), or more (over powered), but we shall consider only properly-powered joints.[14]

The actuator force for an $r$-degree-of-freedom, properly-powered joint is

---

[14]For a different approach to joint force modelling see [53]. $\phi$ and $\psi$ in that paper correspond to $\hat{\mathbf{S}}$ and $\hat{\mathbf{F}}(\hat{\mathbf{S}}^s \hat{\mathbf{F}})^{-1}$ respectively.

given by the $r$-dimensional vector $\mathbf{Q}^a$, and the spatial force produced by the actuation system is given by

$$\hat{\mathbf{f}}^a = \hat{\mathbf{F}}\,\mathbf{Q}^a\,. \tag{7.12}$$

This force is deemed to act on the successor link and be supplied by the predecessor link. The columns of $\hat{\mathbf{F}}$ define the significance of each element of $\mathbf{Q}^a$. The power balance equation now takes on a more complicated form: if $\hat{\mathbf{f}}$ and $\hat{\mathbf{v}}$ are the spatial force and velocity at the joint, then

$$P = \hat{\mathbf{v}}^S\,\hat{\mathbf{f}} = \dot{\mathbf{q}}^T\,\hat{\mathbf{S}}^S\,\hat{\mathbf{F}}\,\mathbf{Q}^a\,. \tag{7.13}$$

To find $\mathbf{Q}^a$ in terms of $\hat{\mathbf{f}}$, use

$$\frac{\partial P}{\partial \dot{\mathbf{q}}} = \hat{\mathbf{S}}^S\,\hat{\mathbf{f}} = \hat{\mathbf{S}}^S\,\hat{\mathbf{F}}\,\mathbf{Q}^a\,,$$

which gives

$$\mathbf{Q}^a = (\,\hat{\mathbf{S}}^S\,\hat{\mathbf{F}}\,)^{-1}\,\hat{\mathbf{S}}^S\,\hat{\mathbf{f}}\,. \tag{7.14}$$

$\hat{\mathbf{S}}^S\,\hat{\mathbf{F}}$ is non-singular, so $\mathbf{Q}^a$ may always be found. If both $\hat{\mathbf{S}}$ and $\hat{\mathbf{F}}$ are constant, then $(\hat{\mathbf{S}}^S\,\hat{\mathbf{F}})^{-1}\,\hat{\mathbf{S}}^S$ can be computed once and its value stored for future use.

$\mathbf{Q}^a$ is a vector of actuation forces for the joint -- it is not a vector of generalised forces. The vector of generalised forces for the joint is given by

$$\mathbf{Q} = \hat{\mathbf{S}}^S\,\hat{\mathbf{F}}\,\mathbf{Q}^a\,.$$

The components of $\mathbf{Q}$ will only be the same as the components of $\mathbf{Q}^a$ if $\hat{\mathbf{S}}^S\,\hat{\mathbf{F}}$ is the identity matrix. In our original model it was assumed that $\hat{\mathbf{F}}$ was a unit pure couple (in the appropriate direction) for a revolute joint and a unit pure force for a prismatic joint. This makes $\hat{\mathbf{S}}^S\,\hat{\mathbf{F}} = 1$. We could choose any value for $\hat{\mathbf{F}}$ which has this property without disturbing the equations of motion.

An actuation force sub-space is not needed for an un-powered joint, since

$\mathbf{Q}^a = \mathbf{0}$ implies $\hat{\mathbf{S}}^S \hat{\mathbf{f}} = \mathbf{0}$ for any permissible value of $\hat{\mathbf{F}}$, which is all we need to know about the spatial force across an un-powered joint. It is sometimes necessary to postulate an active generalised force across an un-powered joint (e.g. when calculating the joint-space inertia matrix), but there is no need for an actuation force.

Having an explicit actuator model allows us to calculate the bearing reaction force from the total force transmitted through a joint. If $\hat{\mathbf{f}}$ is the total joint force, $\hat{\mathbf{f}}^a$ the actuator force, and $\hat{\mathbf{f}}^r$ the bearing reaction force, then

$$\hat{\mathbf{f}}^a + \hat{\mathbf{f}}^r = \hat{\mathbf{f}}\,.$$

Expanding $\hat{\mathbf{f}}^a$ in terms of $\hat{\mathbf{f}}$ using Equations (7.12 and 7.14) gives

$$\hat{\mathbf{f}}^r = (\,\hat{\mathbf{1}} - \hat{\mathbf{F}}\,(\,\hat{\mathbf{S}}^S\,\hat{\mathbf{F}}\,)^{-1}\,\hat{\mathbf{S}}^S\,)\,\hat{\mathbf{f}}\,. \tag{7.15}$$

This equation gives the correct value for the reaction force at a joint provided that $\hat{\mathbf{F}}$ describes the actual forces exerted on the successor link by the actuator.

### 7.4.3. Modifying the Dynamics Algorithms

Multiple-degree-of-freedom joints introduce more substantial changes to the equations of the dynamics algorithms, although there is no change in the procedures for deriving them. The rules for modifying the equations are as follows:

- replace the quantities $\hat{\mathbf{s}}_i$, $\dot{q}_i$ and $\ddot{q}_i$ with $\hat{\mathbf{S}}_i$, $\dot{\mathbf{q}}_i$ and $\ddot{\mathbf{q}}_i$;
- include a term with $\dot{\hat{\mathbf{S}}}_i$ wherever $\hat{\mathbf{S}}_i$ is to be differentiated; and
- replace occurrences of $Q_i$ with the expression $\hat{\mathbf{S}}_i^S\,\hat{\mathbf{F}}_i\,\mathbf{Q}_i^a$.

These rules introduce new terms and factors, and they cause some quantities and expressions which were originally scalars to become vectors or matrices. In some cases the form of the final equations can be quite different, so it is advisable to apply the modification rules to the initial equations and re-derive the final ones.

For example, the link acceleration equation (Equation (4.7)) becomes

$$\hat{\mathbf{a}}_i = \hat{\mathbf{a}}_{i-1} + \hat{\mathbf{v}}_i \times \hat{\mathbf{S}}_i \dot{\mathbf{q}}_i + \hat{\mathbf{S}}_i \ddot{\mathbf{q}}_i + \dot{\hat{\mathbf{S}}}_i \dot{\mathbf{q}}_i \,; \tag{7.16}$$

the equation of motion for the composite-rigid-body method (Equation (5.6)) becomes

$$\mathbf{H}\,\ddot{\mathbf{q}} = \mathbf{K}\,\mathbf{Q}^a - \mathbf{C}\,, \tag{7.17}$$

where $\mathbf{K}$ is a block-wise diagonal matrix with components $\mathbf{K}_{ii} = \hat{\mathbf{S}}_i^S \hat{\mathbf{F}}_i$; the equation for calculating the joint-space inertia matrix (Equation (5.10)) becomes

$$\mathbf{H}_{ji} = \hat{\mathbf{S}}_j^S \hat{\mathbf{I}}_r^C \hat{\mathbf{S}}_i \,; \qquad (\, r = \max\,(i,j)\,) \tag{7.18}$$

and the equation for calculating articulated-body inertias (Equation (6.37)) becomes

$$\hat{\mathbf{I}}_i^A = \hat{\mathbf{I}}_i + \hat{\mathbf{I}}_{i+1}^A - \hat{\mathbf{I}}_{i+1}^A\, \hat{\mathbf{S}}_{i+1} (\hat{\mathbf{S}}_{i+1}^S\, \hat{\mathbf{I}}_{i+1}^A\, \hat{\mathbf{S}}_{i+1})^{-1}\, \hat{\mathbf{S}}_{i+1}^S\, \hat{\mathbf{I}}_{i+1}^A \,. \tag{7.19}$$

In Equation (7.16) there is an extra term involving $\dot{\hat{\mathbf{S}}}_i$. In Equation (7.17) there is an extra factor, $\mathbf{K}$, which transforms the vector of system actuation forces to generalised forces. In Equation (7.18) $\mathbf{H}_{ji}$ has become a rectangular matrix, so now $\mathbf{H}_{ij} = \mathbf{H}_{ji}^T$. ($\mathbf{H}$ is still symmetric and positive definite.) In Equation (7.19) the denominator of the original equation has become an $n_{i+1} \times n_{i+1}$ matrix, which appears as an inverse splitting the original numerator. It is interesting to note that if $n_{i+1}$=6 in this equation then $\hat{\mathbf{S}}_{i+1}^{-1}$ exists and the equation simplifies to $\hat{\mathbf{I}}_i^A = \hat{\mathbf{I}}_i$, which is what you would expect.

The presence of multiple-degree-of-freedom joints makes assessing the computational requirements of the algorithms more complicated than it was before, since some parts depend on the number of joints, and other parts on the number of degrees of freedom of the entire system or of individual joints. For example, the number of link-to-link transformations is a function of the number of joints, the size of the joint-space inertia matrix depends on the total number of degrees of freedom, and the amount of calculation involved in evaluating Equation (7.19) depends on the number of degrees of freedom of joint $i+1$. As a general rule, the computational requirement is more than it would be if each multiple-degree-of-freedom joint were replaced by a single-degree-of-freedom joint, and less than it would be if each were expanded into a kinematically equivalent chain of single-degree-of-freedom joints.

### 7.4.4. More General Constraints

The essential feature of a joint is that it constrains the relative motion of two bodies. An $r$-degree-of-freedom joint imposes $6-r$ constraints. The joints we have considered so far all impose a type of constraint called a scleronomic constraint. This is a constraint on the relative position of two bodies which can be expressed in the form

$$\phi(\mathbf{x}) = \mathbf{0} ,$$

where $\mathbf{x}$ is a 6-dimensional vector of relative position coordinates and $\phi$ is a $6-r$ dimensional vector constraint function. This equation can be solved to express $\mathbf{x}$, and hence the joint displacement, in terms of an $r$-dimensional vector of independent joint variables, $\mathbf{q}$. The joint velocity is a homogeneous linear function of the derivatives of the joint variables, and can be expressed as

$$\hat{\mathbf{v}} = \hat{\mathbf{S}}(\mathbf{q})\, \dot{\mathbf{q}} .$$

$\hat{\mathbf{S}}$ is the $r$-dimensional instantaneous motion sub-space of the joint.

Two other kinds of constraint are possible: rheonomic and non-holonomic. A rheonomic constraint differs from a scleronomic one in that the position constraint is an explicit function of time, i.e.,

$$\phi(\mathbf{x},t) = \mathbf{0} .$$

Both the joint displacement and velocity are functions of time, and the velocity is given by

$$\hat{\mathbf{v}} = \hat{\mathbf{S}}(\mathbf{q},t)\,\dot{\mathbf{q}} + \hat{\xi}(\mathbf{q},t)\,.$$

Rheonomic constraints can be catered for simply by including the extra terms involving $\hat{\xi}$ (and its derivative) in the dynamics equations. Rheonomic and scleronomic constraints are known collectively as holonomic constraints.

Non-holonomic constraints arise when there are additional constraints on the relative velocity of the two bodies. In this case, although the position variables are independent, their derivatives are subject to a set of non-integrable linear constraints $\mathbf{J}\,\dot{\mathbf{q}} = \mathbf{x}$. An example of a non-holonomic constraint is a sphere which is constrained to roll without slipping on a plane surface. The sphere can reach any position on the plane in any orientation, so it has five degrees of position freedom, but because it is not allowed to slip it has only three degrees of velocity freedom. To handle non-holonomic constraints, we must find a set of independent velocity variables, $\mathbf{g}$, which describe the instantaneous relative velocity. $\mathbf{g}$ and $\dot{\mathbf{g}}$ are then used in place of $\dot{\mathbf{q}}$ and $\ddot{\mathbf{q}}$ in the dynamics equations. $\mathbf{q}$ is still needed to describe the position, and its value can be obtained by calculating $\dot{\mathbf{q}}$ from $\mathbf{g}$ and integrating numerically.

## 7.5. Robots with Floating Bases

In our original model it was stipulated that the robot's base member should be stationary, although in fact we can give it any arbitrary predetermined motion without affecting anything other than the starting values for the velocity and acceleration recurrence relations. A robot with a floating base is one whose base member is free to move so that its acceleration depends on the motion of the rest of the robot, and is therefore not known in advance.[15]

The simplest way to allow for a floating base is to treat it as link 1 in a system where joint 1 is a 6-degree-of-freedom un-powered joint connecting link 1 to an inertial reference frame. A partially constrained base can be modelled by giving joint 1 fewer than 6 degrees of freedom, but we will assume that the base has complete motion freedom. If we model a floating base in this way then the forward dynamics can be calculated using the techniques of the previous section.

To model a floating base we shall retain the original numbering scheme, so the base is still link 0, and we shall give it an inertia $\hat{\mathbf{I}}_0$. Its velocity and acceleration are described by $\hat{\mathbf{v}}_0$ and $\hat{\mathbf{a}}_0$ as before, but now $\hat{\mathbf{a}}_0$ is unknown. The external force, if any, acting on the base is $\hat{\mathbf{f}}_0^x$.

The effect of allowing a floating base is different on each of the basic dynamics algorithms. The articulated-body algorithm is the least affected since its articulated bodies are floating structures anyway. The acceleration of the base is given by

$$\hat{\mathbf{a}}_0 = (\hat{\mathbf{I}}_0^A)^{-1} ( \hat{\mathbf{f}}_0^x - \hat{\mathbf{p}}_0 ) , \tag{7.20}$$

where $\hat{\mathbf{I}}_0^A$ and $\hat{\mathbf{p}}_0$ are the articulated-body inertia and bias force of the base, and are calculated from Equations (6.37 and 6.38). This equation takes the place of Equation (6.44). A constant computational cost is incurred which is approximately equal to the incremental cost of one extra joint.

The composite-rigid-body algorithm can be modified as follows. $\mathbf{C}$ is calculated assuming $\hat{\mathbf{a}}_0 = \hat{\mathbf{0}}$. This gives the vector of joint forces needed to account for all dynamic effects except the joint and base accelerations. A force $\hat{\mathbf{f}}_0$ must be applied to the base in order to give it zero acceleration, where

---

[15] These two situations are also known as systems with a kinematic or a dynamic reference frame, respectively [53].

$$\hat{\mathbf{f}}_0 = \hat{\mathbf{f}}_1 + \hat{\mathbf{v}}_0 \hat{\times} \hat{\mathbf{I}}_0 \hat{\mathbf{v}}_0 \,. \tag{7.21}$$

$\hat{\mathbf{f}}_1$ is the force transmitted to link 1, and is obtained during the calculation of $\mathbf{C}$. Ignoring velocity and external effects, the force required at the base to support an acceleration of $\ddot{q}_i$ on joint $i$ is $\hat{\mathbf{I}}_i^C \hat{\mathbf{s}}_i \ddot{q}_i$, and the force required to produce an acceleration of $\hat{\mathbf{a}}_0$ at the base is $\hat{\mathbf{I}}_0^C \hat{\mathbf{a}}_0$; so the equation of motion for the base is

$$\hat{\mathbf{f}}_0^x = \hat{\mathbf{I}}_0^C \hat{\mathbf{a}}_0 + \sum_i \hat{\mathbf{I}}_i^C \hat{\mathbf{s}}_i \ddot{q}_i + \hat{\mathbf{f}}_0$$

$$= \hat{\mathbf{I}}_0^C \hat{\mathbf{a}}_0 + \hat{\mathbf{K}} \ddot{\mathbf{q}} + \hat{\mathbf{f}}_0 \,, \tag{7.22}$$

where $\hat{\mathbf{K}}$ is the $6 \times \mathrm{n}$ matrix whose $i^{th}$ column is $\hat{\mathbf{I}}_i^C \hat{\mathbf{s}}_i$. This gives us six equations in $n$+6 unknowns.

The force required at joint $i$ to compensate for the acceleration of the base is $\hat{\mathbf{s}}_i^S \hat{\mathbf{I}}_i^C \hat{\mathbf{a}}_0$, and the vector of all such forces can be expressed as $\hat{\mathbf{K}}^S \hat{\mathbf{a}}_0$. The equation of motion for the rest of the system is then

$$\hat{\mathbf{K}}^S \hat{\mathbf{a}}_0 + \mathbf{H} \ddot{\mathbf{q}} = \mathbf{Q} - \mathbf{C} \,. \tag{7.23}$$

This gives us another $n$ equations, and we can combine Equations (7.22 and 7.23) to produce the following set of simultaneous equations:

$$\begin{bmatrix} \hat{\mathbf{I}}_0^C & \hat{\mathbf{K}} \\ \hat{\mathbf{K}}^S & \mathbf{H} \end{bmatrix} \begin{bmatrix} \hat{\mathbf{a}}_0 \\ \ddot{\mathbf{q}} \end{bmatrix} = \begin{bmatrix} \hat{\mathbf{f}}_0^x - \hat{\mathbf{f}}_0 \\ \mathbf{Q} - \mathbf{C} \end{bmatrix} \,. \tag{7.24}$$

The augmented matrix is not symmetric, but this is purely because of the way we have arranged the elements in the vectors. Swapping the first three rows of the matrix with the second three will make it symmetric.

The extra calculation required to set up Equation (7.24) is approximately equal to the cost of adding one extra joint in the calculation of the joint-space inertia matrix, but the cost of solving the equations has risen by an amount equal to the addition of six extra degrees of freedom. The incremental cost of a floating base is therefore greater for the composite-rigid-body method than for the articulated-body method.

Inverse dynamics is profoundly affected by the presence of a floating base. The nature of the problem is changed into a hybrid between forward and inverse dynamics: the accelerations of joints $1 \ldots n$ and the force on joint 0 are given, and the task is to find the forces at joints $1 \ldots n$ and the acceleration of joint 0. The algorithm is quite different to the one described in Chapter 4 and is described below in detail. The most general form of hybrid algorithm is one which for each joint computes either the force from the acceleration or the acceleration from the force depending on the value of a flag setting.[16] General hybrid algorithms based on the composite-rigid-body and articulated-body algorithms are also described below.

### 7.5.1. Inverse Dynamics of a Robot with a Floating Base

The base acceleration will be calculated from the composite-rigid-body inertia of the whole robot and the force required to produce zero base acceleration. Once the base acceleration is known, the rest follows easily. Let $\hat{\mathbf{f}}_i$ be the force exerted on link $i$ from link $i-1$, then $\hat{\mathbf{f}}_i$ gives the overall rate of change of momentum for links $i \ldots n$, so

$$\hat{\mathbf{f}}_i = \frac{d}{dt}\left(\sum_{j=i}^{n} \hat{\mathbf{I}}_j \hat{\mathbf{v}}_j\right)$$

$$= \sum_{j=i}^{n} \left(\hat{\mathbf{I}}_j \hat{\mathbf{a}}_j + \hat{\mathbf{v}}_j \hat{\times} \hat{\mathbf{I}}_j \hat{\mathbf{v}}_j\right)$$

[16] Hybrid dynamics can also be viewed as the forward dynamics of a system containing rheonomic constraints.

$$= \left( \sum_{j=i}^{n} \hat{\mathbf{I}}_j \right) \hat{\mathbf{a}}_0 + \sum_{j=i}^{n} \left( \hat{\mathbf{I}}_j \hat{\mathbf{a}}_j^r + \hat{\mathbf{v}}_j \hat{\times} \hat{\mathbf{I}}_j \hat{\mathbf{v}}_j \right)$$

$$= \hat{\mathbf{I}}_i^C \hat{\mathbf{a}}_0 + \hat{\mathbf{p}}_i ; \tag{7.25}$$

where

$$\hat{\mathbf{a}}_i^r = \hat{\mathbf{a}}_i - \hat{\mathbf{a}}_0$$

is the relative acceleration of link $i$ with respect to link 0;

$$\hat{\mathbf{I}}_i^C = \sum_{j=i}^{n} \hat{\mathbf{I}}_j$$

is the composite-rigid-body inertia of links $i \ldots n$; and

$$\hat{\mathbf{p}}_i = \sum_{j=i}^{n} \left( \hat{\mathbf{I}}_j \hat{\mathbf{a}}_j^r + \hat{\mathbf{v}}_j \hat{\times} \hat{\mathbf{I}}_j \hat{\mathbf{v}}_j \right)$$

is the bias force required to give link $i$ zero acceleration. Velocities and relative accelerations are calculated from

$$\hat{\mathbf{v}}_i = \hat{\mathbf{v}}_{i-1} + \hat{\mathbf{s}}_i \dot{q}_i \qquad ( \hat{\mathbf{v}}_0 \text{ known } ) \tag{7.26}$$

and

$$\hat{\mathbf{a}}_i^r = \hat{\mathbf{a}}_{i-1}^r + \hat{\mathbf{s}}_i \ddot{q}_i + \hat{\mathbf{v}}_i \hat{\times} \hat{\mathbf{s}}_i \dot{q}_i , \qquad ( \hat{\mathbf{a}}_0^r = \hat{\mathbf{0}} ) \tag{7.27}$$

composite-rigid-body inertias from

$$\hat{\mathbf{I}}_i^C = \hat{\mathbf{I}}_{i+1}^C + \hat{\mathbf{I}}_i , \qquad ( \hat{\mathbf{I}}_n^C = \hat{\mathbf{I}}_n ) \tag{7.28}$$

and bias forces from

$$\hat{\mathbf{p}}_i = \hat{\mathbf{p}}_{i+1} + \hat{\mathbf{I}}_i \hat{\mathbf{a}}_i^r + \hat{\mathbf{v}}_i \hat{\times} \hat{\mathbf{I}}_i \hat{\mathbf{v}}_i . \qquad ( \hat{\mathbf{p}}_{n+1} = \hat{\mathbf{0}} ) \tag{7.29}$$

The base acceleration is given by

$$\hat{\mathbf{a}}_0 = (\hat{\mathbf{I}}_0^C)^{-1} (\hat{\mathbf{f}}_0^x - \hat{\mathbf{p}}_0) . \tag{7.30}$$

Once $\hat{\mathbf{a}}_0$ is known we can calculate all the $\hat{\mathbf{f}}_i$ from Equation (7.25), so it remains only to calculate the joint forces, which are given by

$$Q_i = \hat{\mathbf{s}}_i^S \hat{\mathbf{f}}_i . \tag{7.31}$$

The computational requirement of this algorithm is considerably more than that of the Newton-Euler algorithm for a robot with a fixed base, but it is less than that of either of the forward dynamics algorithms for robots with a floating base, and its computational complexity is still O $(n)$.

### 7.5.2. Hybrid Dynamics Algorithms

The composite-rigid-body algorithm can be modified to solve problems where the forces at some of the joints and the accelerations of the rest are given. The equation of motion of the system (Equation (5.6)) can be partitioned to give

$$\begin{bmatrix} \mathbf{H}_{11} & \mathbf{H}_{12} \\ \mathbf{H}_{21} & \mathbf{H}_{22} \end{bmatrix} \begin{bmatrix} \ddot{\mathbf{q}}_1 \\ \ddot{\mathbf{q}}_2 \end{bmatrix} = \begin{bmatrix} \mathbf{Q}_1 \\ \mathbf{Q}_2 \end{bmatrix} - \begin{bmatrix} \mathbf{C}_1 \\ \mathbf{C}_2 \end{bmatrix} , \tag{7.32}$$

where $\ddot{\mathbf{q}}_1$ are the unknown accelerations, $\mathbf{Q}_2$ are the unknown forces, and $\ddot{\mathbf{q}}_2$ and $\mathbf{Q}_1$ are the known accelerations and forces. Some swapping of rows and columns may be needed. This equation can be rearranged to put all the unknowns into one vector:

$$\begin{bmatrix} \mathbf{H}_{11} & 0 \\ \mathbf{H}_{21} & -1 \end{bmatrix} \begin{bmatrix} \ddot{\mathbf{q}}_1 \\ \mathbf{Q}_2 \end{bmatrix} = \begin{bmatrix} \mathbf{Q}_1 \\ 0 \end{bmatrix} - \begin{bmatrix} \mathbf{C}_1 + \mathbf{H}_{12} \ddot{\mathbf{q}}_2 \\ \mathbf{C}_2 + \mathbf{H}_{22} \ddot{\mathbf{q}}_2 \end{bmatrix} \tag{7.33}$$

$$= \mathbf{Q}' - \mathbf{C}' .$$

The vector $\mathbf{C}'$ is the vector of joint forces required to give the robot an acceleration of $[\ \mathbf{0}^T\ \ddot{\mathbf{q}}_2{}^T\ ]^T$, and can be calculated from the inverse dynamics in the same way as $\mathbf{C}$. If only the accelerations are of interest, then the equation

$$\mathbf{H}_{11}\ddot{\mathbf{q}}_1 = \mathbf{Q}_1 - (\ \mathbf{C}_1 + \mathbf{H}_{12}\ddot{\mathbf{q}}_2\ ) \tag{7.34}$$

can be used.

The modified algorithm differs from the standard one in the following respects: when calculating $\mathbf{C}$ the unknown accelerations are set to zero and the known ones to their correct values, and when calculating $\mathbf{H}$ only the columns corresponding to the unknown accelerations are needed -- the remaining columns are filled in according to Equation (7.33). If the unknown forces are of no interest then restrict the calculation to just the elements of $\mathbf{C}$ and the rows of $\mathbf{H}$ corresponding to the unknown accelerations.

The articulated-body algorithm can also be modified to solve this kind of problem. For each joint, the equations in Chapter 6 apply if the force is given and the acceleration is to be determined; but if the acceleration is given and the force is to be determined then the equations given below should be used. (Quantities are as defined in Chapter 6.)

The relationship between the force applied to the handle of articulated body $i$ and its acceleration is given by

$$\begin{aligned} \hat{\mathbf{f}} &= \hat{\mathbf{I}}_i^A\,\hat{\mathbf{a}} + \hat{\mathbf{p}}_i \\ &= \hat{\mathbf{I}}_i\,\hat{\mathbf{a}} + \hat{\mathbf{p}}_i^v \\ &\quad + \hat{\mathbf{I}}_{i+1}^A\,(\ \hat{\mathbf{a}} + \hat{\mathbf{v}}_{i+1} \hat{\times} \hat{\mathbf{s}}_{i+1}\,\dot{q}_{i+1} + \hat{\mathbf{s}}_{i+1}\,\ddot{q}_{i+1}\ ) + \hat{\mathbf{p}}_{i+1}\,, \end{aligned} \tag{7.35}$$

so we have that

$$\hat{\mathbf{I}}_i^A = \hat{\mathbf{I}}_i + \hat{\mathbf{I}}_{i+1}^A \tag{7.36}$$

and

$$\hat{\mathbf{p}}_i = \hat{\mathbf{p}}_i^v + \hat{\mathbf{p}}_{i+1} + \hat{\mathbf{I}}_{i+1}^A \left( \hat{\mathbf{v}}_{i+1} \times \hat{\mathbf{s}}_{i+1} \dot{q}_{i+1} + \hat{\mathbf{s}}_{i+1} \ddot{q}_{i+1} \right) . \tag{7.37}$$

Equations (7.36 and 7.37) are used in place of Equations (6.37 and 6.38). Equation (6.41) can be used immediately, since there is no need to calculate $\ddot{q}_i$ first, and the joint force can be calculated from

$$Q_i = \hat{\mathbf{s}}_i^S \left( \hat{\mathbf{I}}_i^A \hat{\mathbf{a}}_i + \hat{\mathbf{p}}_i \right) . \tag{7.38}$$

Equation (7.36) shows that, for the purposes of calculating inertias, if the relative acceleration of two bodies is known then they are effectively rigidly connected.

# Chapter 8
# Coordinate Systems and Efficiency

## 8.1. Introduction

In Chapters 4 to 7 we concentrated on obtaining mathematical formulations for the dynamics algorithms. Implementation details were not considered. The equations which were developed describe in broad terms the computations required to calculate the dynamics on a computer, but the equations by themselves do not constitute complete descriptions of the various algorithms. In particular, three questions have been left unresolved:

- where to put the link coordinate frames,
- how to calculate the link-to-link coordinate transformations, and
- how to implement the operations of spatial arithmetic.

These points have been omitted from earlier discussions since they are largely irrelevant to the mathematical formulations of the algorithms, but they do have an important bearing on the efficiency of the final computer program. The placement of link coordinates and the calculation of transformations are discussed in the next section, and the implementation of spatial arithmetic is discussed in the section after. The final section in this chapter presents the computational requirements of the basic algorithms described in Chapters 4 to 6, and in the case of forward dynamics gives some guidelines as to which algorithm is better under what circumstances.

## 8.2. Link Coordinate Systems

It was stated in the description of the system model in Chapter 4 that a coordinate frame is attached to each link and is used to express the parameters associated with that link -- inertia and inner joint axis. No restrictions were placed on where these coordinate frames should be located. In practice it makes sense to choose locations that simplify some aspect of the computations. There are basically two choices:

- place the origin of the coordinate frame at the link's centre of mass and align the axes with the principal moments of inertia, or

- align the coordinate frame with some suitable feature of the inner joint (e.g., the joint axis).

Adopting the first choice makes the spatial inertia of the link diagonal, which simplifies the computation of the equation of motion. Adopting the second choice makes the link-to-link transformations easier to calculate, and puts the joint axis vector (or sub-space) in a special form which leads to short cuts in several computations. If the dynamics is to be calculated in absolute coordinates then there is no advantage in the first choice, but the second is still useful for its simplified transformations. From the computational point of view, the second choice is more profitable.

### 8.2.1. Denavit-Hartenberg Coordinates

In general, six parameters are needed to specify the transformation from one coordinate frame to another, but with certain classes of mechanism it is possible to position the link coordinate frames so that the link-to-link transformations can be described using fewer parameters. One way to do this is to use the coordinate placement scheme described by Denavit and Hartenberg [15], which uses four parameters to describe the relative positions of adjacent coordinate frames. This scheme is applicable to mechanisms with branch-free connectivity and with joints which are characterised by an axis (revolute, prismatic, cylindric and screw joints). It can be used on any robot mechanism which can be described by the system model of Chapter 4. The scheme makes use of the fact that any two lines in space have a common perpendicular, which is normally unique. The $z$-axes of the link coordinate frames are aligned with the joint axes, and the $x$-axes are aligned with the common perpendiculars between adjacent joint axes.

A coordinate placement based on this scheme is shown in Figure 8-1. $O_i$ is the origin of the coordinate frame attached to link $i$ ($i$-coordinates), which has its $z$-axis aligned with the axis of joint $i$ and its $x$-axis aligned with the common perpendicular between the axes of joints $i$ and $i+1$, both of which are imbedded in link $i$. The transformation from $i$-coordinates to $i+1$ coordinates is effected by a screw transformation on the $x$-axis followed by a screw transformation on the new $z$-axis. The $x$-axis transformation consists of a rotation $\alpha_i$ and a translation $a_i$, and the $z$-axis transformation consists of a rotation $\theta_{i+1}$, a translation $s_{i+1}$ and a rotation or translation of $q_{i+1}$ depending on the type of motion allowed by joint $i+1$. The four parameters $a_i$, $\alpha_i$, $s_{i+1}$ and $\theta_{i+1}$ give the relative position of $i+1$ coordinates with respect to $i$-coordinates when the joint variable $q_{i+1}$ is zero. $O_0{}'$ is the absolute coordinate system (rather than $O_0$) to allow arbitrary placement of the first joint axis, and $O_{n+1}$ is the end effector coordinate system which can be placed arbitrarily with respect to the last joint axis. A total of $4n+6$ parameters define the kinematics of the robot.

The advantage of this scheme is the simple nature of the link-to-link transformations. By using special-purpose screw transformation functions it is possible to reduce the computational cost of performing link-to-link transformations to 80% or less of the cost of a general transformation (see Section 8.3). The disadvantages are that it is not completely general and that a fairly substantial programming effort is required to obtain maximum efficiency. A more general version of this scheme is described in [56].

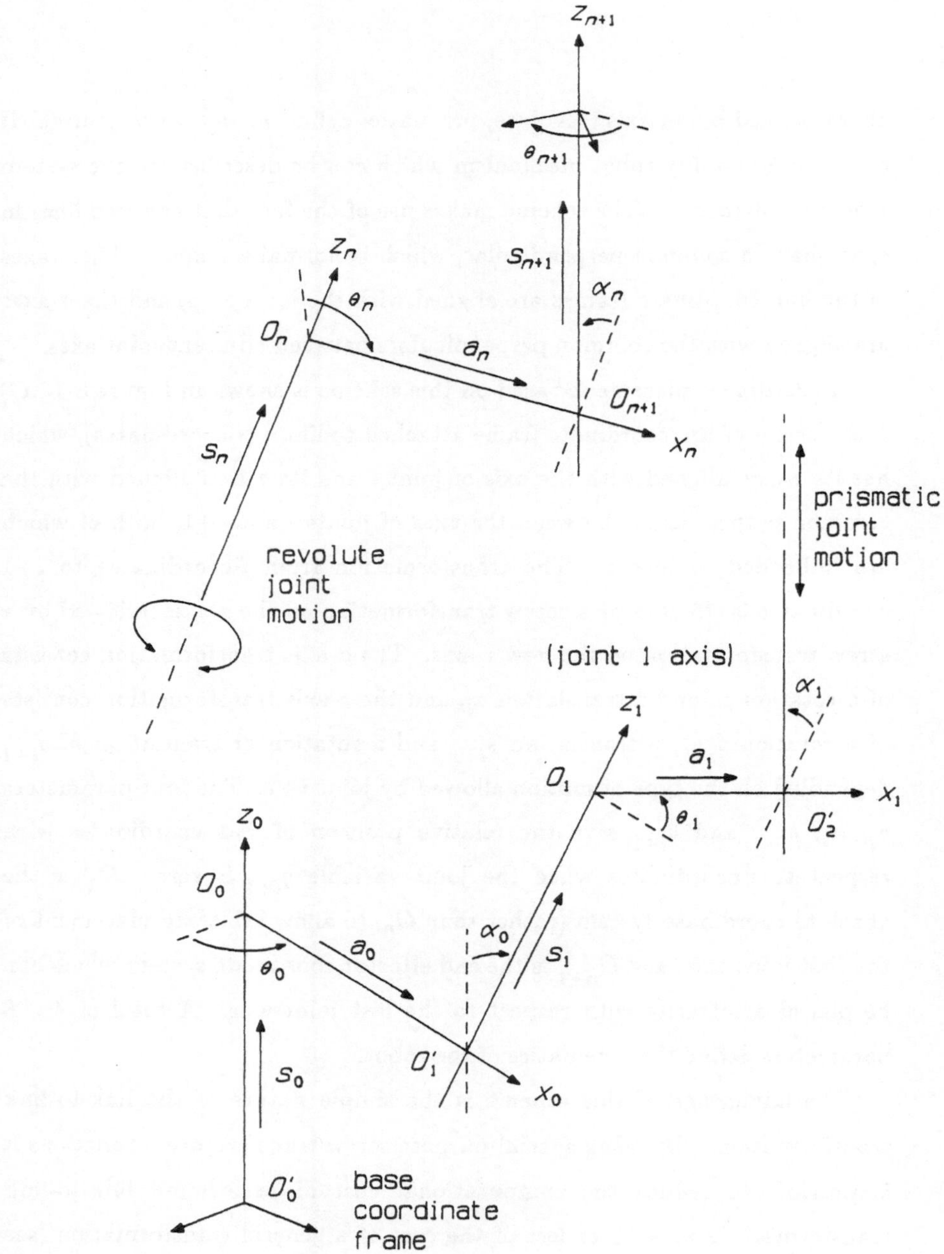


**Figure 8-1:** **Link coordinate systems**

### 8.2.2. Paired Joint Coordinate Frames

The transformation from one link coordinate system to the next consists of a constant part, which reflects the relative positions of two joints imbedded in a single link, and a variable part, which reflects the relative positions of two links connected by a joint. The variable part is a function of the appropriate joint variables, and needs to be recalculated every time the joint's position changes. A good placement of link coordinate frames should minimise the cost of constructing the link-to-link transformations, and it should put the joint axis or sub-space in as convenient a form as possible.

The following scheme is suggested (see Figure 8-2). Two coordinate frames are associated with each joint: an inner frame attached to the link nearer the base (or to the base itself in the case of joint 1) and an outer frame attached to the link farther away from the base. The positions of these frames are chosen to simplify as much as possible the form of the transformation from the inner frame to the outer frame, and to put the representation of the joint's axis or sub-space in the outer coordinate frame in as convenient a form as possible. For example, if a joint is revolute then the two frames are placed so that their origins coincide and their $z$-axes are aligned with the joint axis. The transformation between the two is then a pure rotation about the $z$-axis, and the joint axis is represented by a unit line vector on the $z$-axis.

The link coordinate system for a given link is based on the outer coordinate frame of its inner joint. So, according to our numbering scheme for links and joints, the link coordinate system for link $i$ is based on the outer coordinate frame of joint $i$. The only exception is the base link (link 0), whose coordinates are absolute coordinates. To complete the chain of transformations, an intra-link transformation is defined for each joint, which gives the position of the joint's inner coordinate frame relative to the

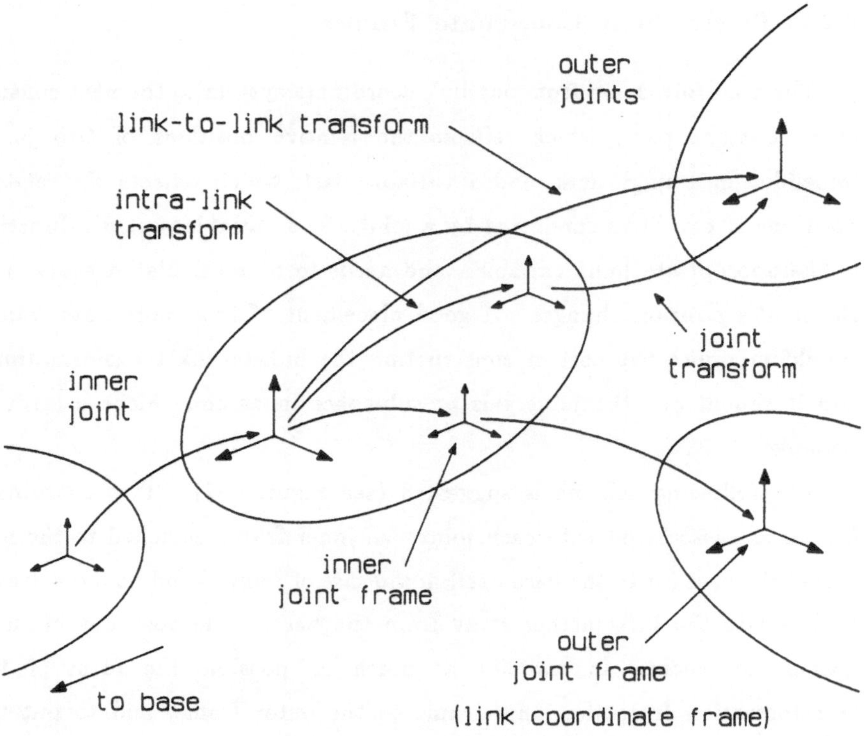


**Figure 8-2:** Paired joint coordinate frames

coordinate system of the preceding link. Each link-to-link coordinate transformation is then given by the product of an intra-link transformation, which is constant, with a joint transformation, which is variable. If we call the coordinate system associated with joint $i$'s inner frame $i'$-coordinates, then the joint transformation is ${}_i\hat{\mathbf{X}}_{i'}$, the intra-link transformation is ${}_{i'}\hat{\mathbf{X}}_{\lambda_i}$, and the link-to-link transformation is ${}_i\hat{\mathbf{X}}_{\lambda_i} = {}_i\hat{\mathbf{X}}_{i'}\,{}_{i'}\hat{\mathbf{X}}_{\lambda_i}$.

Since the link-to-link transformations are general, it is not possible to optimise the operation of applying them to the same extent as with Denavit-Hartenberg coordinates. This is a penalty of increased generality. However, the overall loss of efficiency is small, and the two coordinate placement schemes are equally effective in simplifying the form of joint axis vectors.

| Joint Type | Transformation E | r | Motion Sub-space |
|---|---|---|---|
| revolute | $\begin{bmatrix} c_1 & s_1 & 0 \\ -s_1 & c_1 & 0 \\ 0 & 0 & 1 \end{bmatrix}$ | $\begin{bmatrix} 0 \\ 0 \\ 0 \end{bmatrix}$ | $\begin{bmatrix} 0 \\ 0 \\ 1 \\ 0 \\ 0 \\ 0 \end{bmatrix}$ |
| prismatic | $\begin{bmatrix} 1 & 0 & 0 \\ 0 & 1 & 0 \\ 0 & 0 & 1 \end{bmatrix}$ | $\begin{bmatrix} 0 \\ 0 \\ q_1 \end{bmatrix}$ | $\begin{bmatrix} 0 \\ 0 \\ 0 \\ 0 \\ 0 \\ 1 \end{bmatrix}$ |
| screw (pitch $\rho$) | $\begin{bmatrix} c_1 & s_1 & 0 \\ -s_1 & c_1 & 0 \\ 0 & 0 & 1 \end{bmatrix}$ | $\begin{bmatrix} 0 \\ 0 \\ \rho q_1 \end{bmatrix}$ | $\begin{bmatrix} 0 \\ 0 \\ 1 \\ 0 \\ 0 \\ \rho \end{bmatrix}$ |
| cylindric | $\begin{bmatrix} c_1 & s_1 & 0 \\ -s_1 & c_1 & 0 \\ 0 & 0 & 1 \end{bmatrix}$ | $\begin{bmatrix} 0 \\ 0 \\ q_2 \end{bmatrix}$ | $\begin{bmatrix} 0 & 0 \\ 0 & 0 \\ 1 & 0 \\ 0 & 0 \\ 0 & 0 \\ 0 & 1 \end{bmatrix}$ |

| Joint Type | Transformation E | r | Motion Sub-space |
|---|---|---|---|
| planar | $\begin{bmatrix} c_1 & s_1 & 0 \\ -s_1 & c_1 & 0 \\ 0 & 0 & 1 \end{bmatrix}$ | $\begin{bmatrix} q_2 \\ q_3 \\ 0 \end{bmatrix}$ | $\begin{bmatrix} 0 & 0 & 0 \\ 0 & 0 & 0 \\ 1 & 0 & 0 \\ 0 & 1 & 0 \\ 0 & 0 & 1 \\ 0 & 0 & 0 \end{bmatrix}$ |
| Hooke | $\begin{bmatrix} c_2 & s_1 s_2 & -c_1 s_2 \\ 0 & c_1 & s_1 \\ s_2 & -s_1 c_2 & c_1 c_2 \end{bmatrix}$ | $\begin{bmatrix} 0 \\ 0 \\ 0 \end{bmatrix}$ | $\begin{bmatrix} c_2 & 0 \\ 0 & 1 \\ s_2 & 0 \\ 0 & 0 \\ 0 & 0 \\ 0 & 0 \end{bmatrix}$ |
| spherical | $\begin{bmatrix} c_2 c_3 & c_1 s_3 + s_1 s_2 c_3 & s_1 s_3 - c_1 s_2 c_3 \\ -c_2 s_3 & c_1 c_3 - s_1 s_2 s_3 & s_1 c_3 + c_1 s_2 s_3 \\ s_2 & -s_1 c_2 & c_1 c_2 \end{bmatrix}$ | $\begin{bmatrix} 0 \\ 0 \\ 0 \end{bmatrix}$ | $\begin{bmatrix} c_2 c_3 & s_3 & 0 \\ -c_2 s_3 & c_3 & 0 \\ s_2 & 0 & 1 \\ 0 & 0 & 0 \\ 0 & 0 & 0 \\ 0 & 0 & 0 \end{bmatrix}$ |

**Table 8-1:** Joint Transformations and Axes

### 8.2.3. The Calculation of Joint Transformations

It remains to express the joint transformations and motion sub-spaces in terms of the joint variables, and to decide on positions for the inner and outer joint frames. The simplest rule for placing joint frames is that the axes of the outer frame should be aligned with the rotational and translational freedoms of the joint, and the inner frame should be coincident with the outer frame when the joint variables are zero. Expressions for the joint transformations and motion sub-spaces for a number of common joint types are given in Table 8-1. The transformations are given in terms of a rotation matrix and a translation vector (see Equation (2.18)), and the motion sub-spaces are expressed in the joint's outer coordinate frame. The quantities $q_1$, $q_2$ and $q_3$ are joint variables, and the abbreviations $c_1$, $s_1$, etc. stand for cos $(q_1)$, sin $(q_1)$, etc..

A number of points should be mentioned. Firstly, the axes for revolute, prismatic, screw and cylindric joints are invariant with respect to their joint transforms -- they have the same form in both the inner and outer coordinate frame -- but this is not true of the other joint types. The choice of joint variables for multiple-degree-of-freedom joints affects the form of both the transformation and the motion sub-space. For example, $q_2$ and $q_3$ of the planar joint are the $x$ and $y$ displacements of the joint along the $x$- and $y$-axes of the outer frame. If instead we make them the displacements along the $x$- and $y$-axes of the inner frame then the motion sub-space has the form

$$\hat{S} = \begin{bmatrix} 0 & 0 & 0 \\ 0 & 0 & 0 \\ 1 & 0 & 0 \\ 0 & c_1 & s_1 \\ 0 & -s_1 & c_1 \\ 0 & 0 & 0 \end{bmatrix} ,$$

which is less convenient than the one given in the table. The displacement of the Hooke joint (universal joint) is described by successive rotations of $q_1$ and $q_2$ about the $x$- and $y$-axes; and that of the spherical joint by successive rotations of $q_1$, $q_2$ and $q_3$ about the $x$-, $y$-, and $z$-axes. Where a joint has more than one rotational degree of freedom the motion sub-space matrix is variable, and it is necessary to take account of the terms in the dynamics equations depending on its derivative. This is unavoidable in the case of the Hooke joint, but we can improve the situation with the spherical joint. Finally, note that there is a singularity in the representation of rotations for the spherical joint: if $c_2 = 0$ then the columns of $\hat{\mathbf{S}}$ are linearly dependent, which makes it impossible to calculate $\ddot{\mathbf{q}}$. Numerical problems can be expected in the vicinity of the singularity.

The problems associated with the spherical joint can be alleviated by using a Cartesian angular velocity vector to describe the joint velocity, and Euler parameters to describe the orientation [29]. The velocity across a spherical joint is constrained to be an angular velocity with an axis passing through the rotation centre of the joint, and can be described by a conventional angular velocity vector. If we let $\omega$ be the angular velocity of the outer frame relative to the inner one, expressed in outer frame coordinates, then the form of the motion sub-space relating $\omega$ to the spatial velocity across the joint is

$$\hat{\mathbf{S}} = \begin{bmatrix} 1 & 0 & 0 \\ 0 & 1 & 0 \\ 0 & 0 & 1 \\ 0 & 0 & 0 \\ 0 & 0 & 0 \\ 0 & 0 & 0 \end{bmatrix} .$$

This sub-space is much simpler in form than the one given in the table. It is also constant and singularity-free, and therefore preferable to the one given in the table.

The components of $\omega$ are a better choice for velocity variables than the derivatives of successive rotation angles. The problem is that we can not integrate the angular velocity directly to get the joint's orientation. Fortunately, it is not necessary for the quantities describing the position of a joint to be the integrals of those describing its velocity. It is necessary only that there be some way of calculating changes in position from the velocity.

For our purposes, the most convenient way to describe the orientation of a spherical joint is to use Euler parameters, which are a set of four quantities, $q_0$, $q_1$, $q_2$ and $q_3$, satisfying a normality constraint $q_0^2+q_1^2+q_2^2+q_3^2=1$. The Euler parameters are the elements of a unit quaternion $q$ describing a half-angle rotation [50]. The rate of change of orientation can be calculated from the differential equation

$$\begin{bmatrix} \dot{q}_0 \\ \dot{q}_1 \\ \dot{q}_2 \\ \dot{q}_3 \end{bmatrix} = 1/2 \begin{bmatrix} -q_1 & -q_2 & -q_3 \\ q_0 & -q_3 & q_2 \\ q_3 & q_0 & -q_1 \\ -q_2 & q_1 & q_0 \end{bmatrix} \begin{bmatrix} \omega_1 \\ \omega_2 \\ \omega_3 \end{bmatrix} \tag{8.1}$$

(which can be expressed in quaternion algebra as $\dot{q} = 1/2\, q\, \omega$). The constraint can be maintained by normalising the Euler parameters periodically. The $3 \times 3$ rotation matrix of the joint transformation is given in terms of $q$ by

$$\mathrm{E} = 2 \begin{bmatrix} q_0^2+q_1^2-1/2 & q_1q_2+q_0q_3 & q_1q_3-q_0q_2 \\ q_1q_2-q_0q_3 & q_0^2+q_2^2-1/2 & q_2q_3+q_0q_1 \\ q_1q_3+q_0q_2 & q_2q_3-q_0q_1 & q_0^2+q_3^2-1/2 \end{bmatrix} . \tag{8.2}$$

The advantages of this approach are that there are no singularities in the representation of orientations and that the joint sub-space has a simpler form which is constant in link coordinates. The disadvantages are that we have more variables, a constraint equation to maintain, and a differential equation in place of simple integration. On balance, the advantages outweigh the disadvantages.

## 8.3. Efficient Spatial Arithmetic

One way to implement the dynamics algorithms described in previous chapters is to rewrite the spatial equations in terms of the linear and angular components of the spatial quantities, and use standard 3-dimensional vector and matrix arithmetic functions to perform the calculations. The resulting computer code executes efficiently, but it is rather lengthy compared with the spatial equations specifying the algorithms. If the translation can be done automatically then this is a good option, since the programmer can specify the algorithms concisely in spatial notation and leave to the computer the task of producing efficient code; but if the translation has to be done manually then the process can be lengthy, tedious and error prone, and errors in the expanded code may be difficult to track down.

Alternatively, we can implement the dynamics algorithms directly in terms of calls to spatial arithmetic functions. This avoids the need to do a translation, but it introduces the need to construct a spatial arithmetic package. The simplest approach is to take a standard matrix arithmetic package and augment it with functions to do spatial inner, outer and cross products. The spatial quantities are represented by $6 \times 1$ and $6 \times 6$ matrices. The computer code for the dynamics algorithms is now much shorter, but it is also highly inefficient. The inefficiency stems primarily from the use of $6 \times 6$ matrices, which have 36 elements, to represent rigid-body transformations, which have only six independent parameters. The redundancy in this representation leads to considerable inefficiencies in storing, combining and applying transformations [51]. The representation of rigid-body inertias, which have ten independent parameters, is also inefficient.

To improve the efficiency of the spatial arithmetic package it is necessary to adopt compact computer representations for spatial quantities and write a suite of special-purpose arithmetic functions to operate on these compact representations. This combines the efficiency of the first approach with the convenience of the second. Its main disadvantage is the increased effort involved in creating the spatial arithmetic package in the first place.

### 8.3.1. Computer Representation of Spatial Quantities

To achieve efficient spatial arithmetic, certain spatial quantities have to be broken down into linear and angular components which can be represented more economically on a computer, and the arithmetic operations should be performed using these components. This can be done if spatial quantities are represented as collections or aggregates of components. Most programming languages will allow such aggregates to be treated as objects in their own right.

| Quantity | Mathematical Form | Computer Representation | No. of Real Numbers Math. | Comp. |
|---|---|---|---|---|
| vector | $\begin{bmatrix} \mathbf{v} \\ \mathbf{v}_O \end{bmatrix}$ | $spv(\mathbf{v},\mathbf{v}_O)$ | 6 | 6 |
| transform | $\begin{bmatrix} \mathbf{E} & \mathbf{0} \\ \mathbf{E}\mathbf{r}\times^T & \mathbf{E} \end{bmatrix}$ | $spx(\mathbf{E},\mathbf{r})$ | 36 | 12 |
| artic.-body inertia | $\begin{bmatrix} \mathbf{H}^T & \mathbf{M} \\ \mathbf{I} & \mathbf{H} \end{bmatrix}$ | $abi(\mathbf{M},\mathbf{H},\mathbf{I})$ | 36 | 27 |
| rigid-body inertia | $\begin{bmatrix} \mathbf{h}\times^T & m\mathbf{1} \\ \mathbf{I} & \mathbf{h}\times \end{bmatrix}$ | $rbi(m,\mathbf{h},\mathbf{I})$ | 36 | 13 |

**Table 8-2:** Computer representation of spatial quantities

The suggested computer representations for spatial quantities are given in Table 8-2. The notation *obj*(*a*, *b*, . . . ) means the spatial quantity of type *obj* which can be constructed from the quantities listed. The computer representation is an array which is the concatenation of the numbers making up the quantities in the list. So, for example, a spatial rigid-body inertia is stored as a 13 element array, the first element being the mass, the next three being the elements of **h**, and the remainder being the elements of **I**. Operations like addition and multiplication by a scalar can be implemented on the computer representations in a straightforward manner, for example:

$$spv(\,\mathbf{v}_1, \mathbf{v}_{O1}\,) + spv(\,\mathbf{v}_2, \mathbf{v}_{O2}\,) = spv(\,\mathbf{v}_1{+}\mathbf{v}_2, \mathbf{v}_{O1}{+}\mathbf{v}_{O2}\,)\,,$$

$$\alpha \times rbi(\,m, \mathbf{h}, \mathbf{I}\,) = rbi(\,\alpha m, \alpha\mathbf{h}, \alpha\mathbf{I}\,)\,,$$

$$rbi(\,m, \mathbf{h}, \mathbf{I}_1\,) + abi(\,\mathbf{M}, \mathbf{H}, \mathbf{I}_2\,) = abi(\,m\mathbf{1}{+}\mathbf{M}, \mathbf{h}{\times}{+}\mathbf{H}, \mathbf{I}_1{+}\mathbf{I}_2\,)\,.$$

The first two can be implemented using standard matrix arithmetic functions, but the third is better done by a special-purpose function. Multiplication of inertias by vectors and the various vector products are also fairly straightforward, although some of the transformations are a little complicated.

### 8.3.2. Efficient Spatial Transformations

Spatial transformations account for a large part of the computational cost of the dynamics algorithms (especially in link coordinates), so it is worthwhile considering how they can be implemented efficiently. Let $\hat{\mathbf{X}}$, $\hat{\mathbf{X}}_1$ and $\hat{\mathbf{X}}_2$ be spatial transformations defined by

$$\mathbf{X} = \begin{bmatrix} \mathbf{E} & \mathbf{0} \\ \mathbf{E}\,\mathbf{r}{\times}^T & \mathbf{E} \end{bmatrix}, \quad \hat{\mathbf{X}}_1 = \begin{bmatrix} \mathbf{E}_1 & \mathbf{0} \\ \mathbf{E}_1\mathbf{r}_1{\times}^T & \mathbf{E}_1 \end{bmatrix}$$

and

$$\hat{\mathbf{X}}_2 = \begin{bmatrix} \mathbf{E}_2 & \mathbf{0} \\ \mathbf{E}_2\mathbf{r}_2\times^T & \mathbf{E}_2 \end{bmatrix}.$$

Given that $\hat{\mathbf{X}} = \hat{\mathbf{X}}_1 \hat{\mathbf{X}}_2$, we have that

$$\begin{bmatrix} \mathbf{E} & \mathbf{0} \\ \mathbf{E}\mathbf{r}\times^T & \mathbf{E} \end{bmatrix} = \begin{bmatrix} \mathbf{E}_1 & \mathbf{0} \\ \mathbf{E}_1\mathbf{r}_1\times^T & \mathbf{E}_1 \end{bmatrix} \begin{bmatrix} \mathbf{E}_2 & \mathbf{0} \\ \mathbf{E}_2\mathbf{r}_2\times^T & \mathbf{E}_2 \end{bmatrix}$$

$$= \begin{bmatrix} \mathbf{E}_1\mathbf{E}_2 & \mathbf{0} \\ \mathbf{E}_1\mathbf{r}_1\times^T\mathbf{E}_2 + \mathbf{E}_1\mathbf{E}_2\mathbf{r}_2\times^T & \mathbf{E}_1\mathbf{E}_2 \end{bmatrix}$$

$$= \begin{bmatrix} \mathbf{E}_1\mathbf{E}_2 & \mathbf{0} \\ \mathbf{E}_1\mathbf{E}_2(\mathbf{E}_2^{-1}\mathbf{r}_1 + \mathbf{r}_2)\times^T & \mathbf{E}_1\mathbf{E}_2 \end{bmatrix}.$$

So, in terms of the computer representations of the transformations, we have that

$$spx(\,\mathbf{E}_1, \mathbf{r}_1\,)\; spx(\,\mathbf{E}_2, \mathbf{r}_2\,) = spx(\,\mathbf{E}_1\mathbf{E}_2, \mathbf{r}_2 + \mathbf{E}_2^{-1}\mathbf{r}_1\,)\,.$$

The implementation of this operation requires one $3 \times 3$ matrix multiplication, one transpose, one matrix-vector multiplication and one vector addition. The total computational requirement is $36m+27a$, where $m$ stands for scalar multiplication and $a$ for scalar addition. This compares favourably with the computational requirement for a $6 \times 6$ matrix multiplication, which is $216m+180a$. Other operations involving transformations may be treated in a similar way, and the results are summarised in Table 8-3. (It is assumed that the operations $\mathbf{a}\times\mathbf{b}\times$ and $\mathbf{a}\times\mathbf{B}$ take $9m+3a$ and $18m+9a$ respectively.) The figures in the improvement column indicate the factor by which the computational cost is

| Operation | Result | Cost | Improvement |
|---|---|---|---|
| $spx(\mathbf{E}_1,\mathbf{r}_1)\ spx(\mathbf{E}_2,\mathbf{r}_2)$ | $spx(\mathbf{E}_1\mathbf{E}_2,\ \mathbf{r}_2+\mathbf{E}_2^{-1}\mathbf{r}_1)$ | $36m+27a$ | 6 |
| $spx(\mathbf{E},\mathbf{r})\ spv(\mathbf{v},\mathbf{v}_O)$ | $spv(\mathbf{E}\mathbf{v},\ \mathbf{E}(\mathbf{v}_O-\mathbf{r}\times\mathbf{v}))$ | $24m+18a$ | 1.5 |
| $spx(\mathbf{E},\mathbf{r})\ abi(\mathbf{M},\mathbf{H},\mathbf{I})$ | $abi(\mathbf{E}\mathbf{M}\mathbf{E}^{-1},\ \mathbf{E}(\mathbf{H}-\mathbf{r}\times\mathbf{M})\mathbf{E}^{-1},$ $\mathbf{E}(\mathbf{I}-\mathbf{r}\times\mathbf{H}^T+(\mathbf{H}-\mathbf{r}\times\mathbf{M})\mathbf{r}\times)\mathbf{E}^{-1})$ | $216m+162a$ | 2 |
| $spx(\mathbf{E},\mathbf{r})\ rbi(m,\mathbf{h},\mathbf{I})$ | $rbi(m,\ \mathbf{E}(\mathbf{h}-m\mathbf{r}),$ $\mathbf{E}(\mathbf{I}+\mathbf{r}\times\mathbf{h}\times+(\mathbf{h}-m\mathbf{r})\times\mathbf{r}\times)\mathbf{E}^{-1})$ | $84m+69a$ | 5 |

**Table 8-3:** Implementation of transformation operations

reduced over the use of 6-dimensional vectors and matrices.[17] The improvement in efficiency is quite substantial.

### 8.3.3. Symmetry

The process of reducing the computational requirement can be taken much further. For instance, we can make use of the fact that **M** and **I** are symmetric, and implement special functions for the addition and multiplication of matrices where the result is known to be symmetric. (Symmetric matrices can be stored in a compact form, but this is not recommended here since it is more trouble than it is worth.) The resulting improvement in efficiency is about 20%, which is quite small, and the numbers are given in Table 8-4.

[17] In most cases both multiplication and addition requirements are reduced by roughly the same factor, but where not the improvement figure is calculated assuming multiplication takes 4 times as long as addition.

| Operation | Cost before | Cost after | Improvement |
|---|---|---|---|
| *spx*(E,r) *abi*(M,H,I) | $216m+162a$ | $177m+132a$ | 1.2 |
| *spx*(E,r) *rbi*($m$,h,I) | $84m+69a$ | $69m+57a$ | 1.2 |

**Table 8-4:** Making use of symmetry

### 8.3.4. Screw Transformations on Coordinate Axes

Recall that Denavit-Hartenberg coordinates allow the transformation from one link coordinate system to an adjacent one to be performed by two screw transformations, one on the $x$-axis and one on the $z$-axis. When the screw axis is one of the coordinate axes, the transformation takes on a particularly simple form (e.g., see [51]) which will be represented as *scr*($\theta$, $d$, $ax$). $\theta$ is the angle of rotation about the axis, $d$ the displacement along it, and $ax$ the specification of which axis ($x$, $y$, or $z$) is involved. For example, *scr*($\theta$, $d$, $z$) is equivalent to *spx*(**E**, **r**) where

$$\mathbf{E} = \begin{bmatrix} c & s & 0 \\ -s & c & 0 \\ 0 & 0 & 1 \end{bmatrix}$$

and

$$\mathbf{r} = \begin{bmatrix} 0 \\ 0 \\ d \end{bmatrix},$$

the abbreviations $s$ and $c$ standing for sin ($\theta$) and cos ($\theta$) respectively. The application of *scr*($\theta$, $d$, $z$) to *spv*(v, $\mathrm{v}_O$) proceeds as follows:

$$\begin{bmatrix} c & s & 0 & 0 & 0 & 0 \\ -s & c & 0 & 0 & 0 & 0 \\ 0 & 0 & 1 & 0 & 0 & 0 \\ -sd & cd & 0 & c & s & 0 \\ -cd & -sd & 0 & -s & c & 0 \\ 0 & 0 & 0 & 0 & 0 & 1 \end{bmatrix} \begin{bmatrix} v_x \\ v_y \\ v_z \\ v_{Ox} \\ v_{Oy} \\ v_{Oz} \end{bmatrix} = \begin{bmatrix} cv_x+sv_y \\ cv_y-sv_x \\ v_z \\ c(v_{Ox}+v_yd)+s(v_{Oy}-v_xd) \\ c(v_{Oy}-v_xd)-s(v_{Ox}+v_yd) \\ v_{Oz} \end{bmatrix}$$

The computational requirement for this operation is $10m+6a$, which is less than half that of the general transformation, so it is possible to save time by performing transformations between adjacent coordinate systems as pairs of these transformations.

| Operation | Cost | Improvement |
|---|---|---|
| $scr(\theta,d,ax)\ spx(\mathbf{E},\mathbf{r})$ | $15m+9a$ | 1.25 |
| $scr(\theta,d,ax)\ spv(\mathbf{v},\mathbf{v}_O)$ | $10m+6a$ | 1.25 |
| $scr(\theta,d,ax)\ abi(\mathbf{M},\mathbf{H},\mathbf{I})$ | $70m+64a$ | 1.5 |
| +symmetry | $54m+50a$ | 1.6 |
| $scr(\theta,d,ax)\ rbi(m,\mathbf{h},\mathbf{I})$ | $31m+25a$ | 1.35 |
| +symmetry | $23m+20a$ | 1.5 |

**Table 8-5:** Coordinate-axis-screw transformations

The computational requirements for the operations of table 8-3 using coordinate-axis-screw transformations are given in table 8.5. The figures in the cost column refer to one transformation, but those in the improvement column refer to using two coordinate-axis-screw transformations instead of one general one. To save having to compute trigonometric functions every time a transformation is performed, the aggregate describing the screw transformation should include the values $s=\sin(\theta)$ and $c=\cos(\theta)$, and for transforming matrices it is assumed that $s^2$, $c^2$ and $sc$ are also available (otherwise an additional $3m$ overhead per transformation is incurred). Notice that the combination of coordinate-axis-screw transformations and symmetric matrices is more effective than either separately. An additional bonus with this method is that we do not have the overhead of constructing explicitly the transformations between adjacent link coordinate systems.

Even this is not the last word in optimisation, for if we take the operation $scr(\theta, d, ax)\ rbi(m, \mathbf{h}, \mathbf{I})$ and find an expression for each element of the result, then, by judicious use of partial results, we can reduce the computational cost to a mere $19m+18a$.

### 8.3.5. Parallel Computation

It has been assumed throughout the preceding discussion that the calculations are to be performed on a serial computer. Suppose we have a computer capable of performing three arithmetic operations in parallel so that, in terms of the time taken, the cost of a vector addition is reduced from $3a$ to $a$, that of a scalar product from $3m+2a$ to $m+a$, and that of a vector product from $6m+3a$ to $2m+a$; but the hardware is such that we can not ask for three arbitrary scalar operations to be performed in parallel. This gives 3-dimensional vector operations an advantage over scalar operations with the result that the optimisations based on symmetric matrices and coordinate-axis-screw transformations no longer give any improvement in computation speed. However, the aggregate representation retains its margin of efficiency over the 6-dimensional vector and matrix representation. On a machine capable of performing six arithmetic operations in parallel, the aggregate representation is still faster, though with a reduced margin of efficiency. On a machine capable of performing 36 or more operations in parallel it is better to use the 6-dimensional vector representation.

## 8.4. The Computational Requirements of the Algorithms

This section presents the computational requirements of each of the three basic dynamics algorithms and compares the solution of the forward dynamics problem via the composite-rigid-body method with that via the articulated-body method. The computational requirements of the extended algorithms described in Chapter 7 are harder to assess, since they depend on more parameters, and will not be considered here.

The tally of arithmetic operations is a good guide to the actual amount of effort involved in the computation process. The assumption is that these operations (especially multiplication) take significantly longer to execute than the average machine instruction, and so account for most of the execution time. The computational requirement for a particular algorithm depends on the degree of optimisation, and to compare the relative performance of algorithms requires that they be implemented fairly. Here it will be assumed that Denavit-Hartenberg coordinates are used, and that all of the optimisation techniques described in the previous section are being used. Optimisations on operations other than transformations are also taken into account, and these are described below along with a discussion on the merits of calculation in link and absolute coordinates.

### 8.4.1. Link Coordinates versus Absolute Coordinates

To compute the algorithms in absolute coordinates requires the overhead of computing $\hat{\mathbf{I}}_i$ and $\hat{\mathbf{s}}_i$ from $\hat{\mathbf{I}}_i'$ and $\hat{\mathbf{s}}_i'$ (see Chapter 4, Section 4). This involves the calculation of the $n$ transformations ${}_0\hat{\mathbf{X}}_i$ and performing $n$ vector and $n$ rigid-body inertia transformations. To compute the algorithms in link coordinates requires that every passed-on quantity be transformed to an adjacent coordinate system. In terms of the number of quantities that need to be transformed in each case, it would seem that absolute coordinates are more efficient. However, link coordinates have a number of points in their favour:

- a transformation to an adjacent coordinate system can be made more efficient than a general transformation,
- $\hat{\mathbf{s}}_i'^S$ can be made to have a special form in link coordinates which leads to considerable savings, and
- there is no overhead for calculating ${}_0\hat{\mathbf{X}}_i$.

The second point is based on the fact that $\hat{\mathbf{s}}_i'^S$ can be made equal to one of the coordinate axes, so much multiplication can be avoided on operations like $Q_i = \hat{\mathbf{s}}_i'^S \hat{\mathbf{f}}_i$ (Equation (4.15)), which simplifies to selecting an element from $\hat{\mathbf{f}}_i$, and $\hat{\mathbf{h}}_i = \hat{\mathbf{I}}_i^A \hat{\mathbf{s}}_i'$ (Equation (6.57)), which simplifies to selecting a column (or row) from $\hat{\mathbf{I}}_i^A$. The third point is only valid if the locations of the links are of no interest.

A disadvantage of performing the calculation in link coordinates is that if external forces are being applied to the links then they must be transformed to link coordinates before their effects can be taken into account. This involves an additional transformation overhead which is not incurred if the calculations are being performed in absolute coordinates.

In the case of the Newton-Euler algorithm, computation in link coordinates has the considerable advantage of not requiring the transformation of any tensors, and is therefore substantially more efficient than computation in absolute coordinates.

In the case of the articulated-body algorithm, computation in link coordinates requires the transformation of articulated-body inertias, which is expensive, whereas computation in absolute coordinates requires the transformation of rigid-body inertias, which is somewhat cheaper. Link coordinates are still faster, but only by about 10% or so, and the need to take into account external forces on the links or the location of the end effector can easily render link coordinates less efficient.

In the case of the composite-rigid-body algorithm the situation is more complicated, since the link coordinate version requires $O(n^2)$

transformations whereas the absolute coordinate version requires only O $(n)$. As a result, the absolute coordinate version has a considerably lower coefficient of $n^2$ in its computational requirement, but because of the overheads associated with the use of absolute coordinates it has a considerably larger O $(n)$ dependency. The absolute coordinate version is more efficient for values of $n > 16$, but its O $(n)$ dependency is so large that solving forward dynamics by this method is always less efficient than the articulated-body algorithm.

| **Method** | **Cost** | | **Cost** |
|---|---|---|---|
| | **multiplications** | **additions** | $n=6$ |
| Newton-Euler | $130n-68$ | $101n-56$ | $712m+550a$ |
| composite-rigid-body | $10n^2+31n-41$ | $6n^2+40n-46$ | $505m+410a$ |
| fwd. dyn. via CRBM | $1/6n^3+11\ 1/2n^2+160\ 1/3n-109$ | $1/6n^3+7n^2+138\ 5/6n-102$ | $1303m+1019a$ |
| articulated body | $300n-267$ | $279n-259$ | $1533m+1415a$ |

**Table 8-6:** Comparison of dynamics algorithms

### 8.4.2. Comparison of the Algorithms

Table 8-6 shows the computational requirements of the three algorithms in terms of the numbers of arithmetic operations expressed as polynomials in $n$. The numbers exclude the cost of calculating the $n-1$ sines and cosines required for the transformations. The computational requirement of performing the complete forward dynamics calculation via the composite-rigid-body method is also included for comparison with the articulated-body method. This is the sum of the computational requirements of the composite-rigid-body method, a simplified version of the Newton-Euler method and the $L\,D\,L^T$ decomposition method for solving a set of simultaneous equations where the coefficient matrix is symmetric and positive definite. In all cases the algorithms are performed in link coordinates using all the optimisations described above, but the numbers in the table do not represent the minimum possible cost. They do, however, give a reasonably fair comparison of the relative costs of the algorithms.

The basic conclusions we can draw are that forward dynamics takes longer to compute than inverse dynamics, by a factor of up to 2.3, and that the articulated-body method for forward dynamics is only faster than the composite-rigid-body method if $n > 9$, so the composite-rigid-body method is better for most practical problems. If we allow 200$\mu$s computation time for a sine/cosine calculation, 10$\mu$s for multiplication, 2$\mu$s for addition and a 1$\mu$s overhead per operation for shuffling data, we get an execution time of less than 12ms for the Newton-Euler algorithm on a 6-degree-of-freedom robot, which is fast enough for use in real-time control.

Various other factors may tip the balance in favour of one or another algorithm. For example, if the robot has a floating base (e.g., as in a satellite-mounted robot) then the articulated-body algorithm becomes relatively much more efficient than the composite-rigid-body algorithm (see Chapter 7), but the presence of kinematic loops is detrimental to the articulated-body algorithm (see Chapter 9). Another factor that should be taken into account is the possibility of partitioning the calculation of forward dynamics into parts that need frequent re-calculation and parts that do not (as done in [19] for example). This practice favours the composite-rigid-body method.

### 8.4.3. Approximations and Special-Purpose Algorithms

All the discussion so far has been on the implementation and optimisation of exact dynamics algorithms for robots with general geometries. Most industrial robots have very special geometries, with adjacent joint axes being almost always parallel or perpendicular and frequently intersecting. It is possible to write special-purpose algorithms which take advantage of the special features of some particular robot geometry to reduce the computational cost. It is also possible to generate such algorithms automatically. Another way to reduce the cost is to use approximations to the inertia parameters. A link's centre of mass is typically close to a link coordinate axis, and there is usually little error in assuming that its principal axes of inertia are parallel with the link coordinate axes. The extent of the reduction in computational cost over the general algorithms depends on the particular robot, but typically a factor of 2 or more can be expected. For some examples see [24], [32], [34], [47], [72].

# Chapter 9
# Contact, Impact, and Kinematic Loops

## 9.1. Introduction

This chapter is concerned with the motion simulation (i.e., forward dynamics) of robot mechanisms containing kinematic loops, and robots in contact with rigid bodies in their environment. Allowing for kinematic loops is the last step in the process of generalising the robot mechanism; while taking account of contact between robots and workpieces is the first step in simulating the interaction between robots and their environment. The two problems are similar in that in each case we are imposing additional motion constraints on a mechanism, and in each case the problem is solved by formulating the equation of motion of an unconstrained mechanism subject to unknown constraint forces. The two problems differ in the nature of the constraints they impose.

A kinematic loop occurs in a mechanism if the connectivity is such that a circuit can be traced from one link back to itself without traversing any joint more than once. Kinematic loops occur in many robot mechanisms, typically as a means of actuation, and they are almost universal in non-robotic mechanisms. A great deal of work has been done on the motion simulation of mechanisms with kinematic loops; e.g., [45], [58], [59], [66], [67], [44], [76], [54].

Kinematic loops introduce two complications: firstly, they introduce constraint equations (called loop equations) between the joint variables, so there is no longer a one-to-one correspondence between the joint variables and the motion freedoms of the system; and secondly, they introduce a number of unknown reaction forces, the number of forces being equal to the number of motion constraints. The approach that will be used to obtain the equations of motion for a mechanism containing kinematic loops is to replace the loop-closing joints with unknown reaction forces acting on an open-loop mechanism, find the equations of motion for this mechanism, and augment the equations with the acceleration constraint equations imposed by the loops. This method is essentially the same as that described in [58] and the method of Lagrange multipliers described in [45], although these papers deal with planar mechanisms. An algorithm using the articulated-body approach is also described. This algorithm is not very efficient, and is included mainly for interest.

A 'contact' is a kinematic constraint which arises from physical contact between rigid bodies which are capable of separating. The condition for two contacting bodies to separate depends on the forces acting on them. Although joints also arise from contact between rigid bodies, it is normally assumed that the two bodies linked by a joint can not, or do not, separate.

The dynamics of contacting bodies is more complicated than that of bodies connected by joints. This is because the motion constraints can only be formulated as non-linear functions of the internal forces in the system, which themselves depend on the motion constraints. The method used to solve for the motion of such systems is to find the equation of motion of the system without contacts, but subject to unknown contact forces. The equation of motion is then substituted into the constraint equations to give a set of simultaneous quadratic equations in the unknown contact forces. The root which satisfies a linear inequality is the set of contact forces, which can be substituted into the equation of motion to find the accelerations. It is assumed that all contacts are point contacts, and the effects of friction between contacting bodies are ignored.

Contacts can be made as well as broken. The creation of a new contact is usually the result of a collision between two bodies. At the moment of collision, the velocities of the bodies do not satisfy the new motion constraints, and impulses must be applied to adjust the velocities. The phase of motion during which impulses are acting on the system will be called impact. The equations of impulsive motion for systems of rigid bodies are similar to the equations of continuous motion. The detection of collisions between moving bodies is essentially a geometric problem, and is outside the scope of this book.

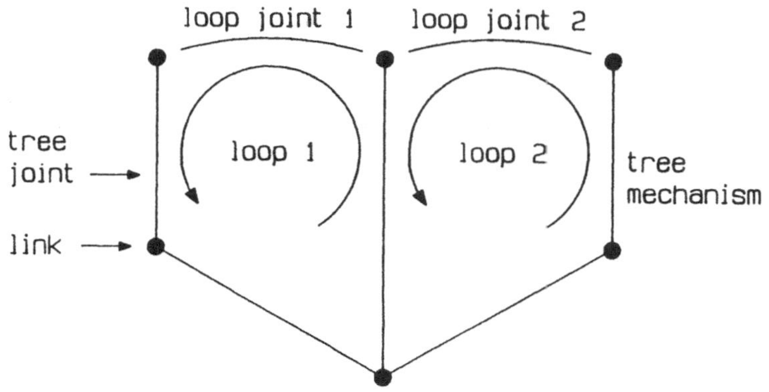


**Figure 9-1:** Mechanism with kinematic loops

## 9.2. Kinematic Loops

A kinematic loop exists in a mechanism if it is possible to trace a circuit from some link back to itself without traversing any joint more than once. Kinematic loops can be introduced into a mechanism by adding extra joints. Each such joint closes one independent kinematic loop. For example, Figure 9-1 shows the connection graph of a tree-structure mechanism to which two extra joints have been added. The nodes in this graph represent the links in the mechanism, and the arcs represent the joints connecting them. The two extra joints and the loops they close are indicated on the diagram. There is in fact a third loop in this mechanism, but this loop is dependent on the other two and may be ignored.

Any mechanism containing kinematic loops can be constructed from a tree-structure mechanism by adding joints; and conversely, any mechanism containing loops can be converted to a tree-structure mechanism by cutting certain joints. A mechanism with $N$ links (excluding the base) and $L$ independent kinematic loops has $N+L$ joints: $N$ joints incorporated in the tree and $L$ accounting for the loops. There is some choice as to which joints in the mechanism are to be treated as loop joints, but the total number is fixed.

A robot mechanism containing kinematic loops will be modelled as a tree-structure mechanism plus a list of extra joints which close the loops. These joints will be called loop joints to distinguish them from those in the tree, which will be called tree joints. For each loop joint it is necessary to specify which link is its predecessor and which its successor, and additional intra-link transformations are needed to locate the joint's inner and outer coordinate frames in these links. The loop joints will be numbered $1\,..\,L$, where $L$ is the number of loops, and the kinematic loop closed by joint $k$ will be called loop $k$. The number of tree joints in the mechanism and the number of degrees of freedom of the tree will be denoted by $N$ and $n$ respectively. Multiple-degree-of-freedom joints will be assumed.

### 9.2.1. The Effect of Kinematic Loops on Mechanism Dynamics

Let us consider the effect of a loop joint on the dynamics of a tree-structure mechanism. Suppose that joint $k$ closes a kinematic loop on a tree-structure mechanism by connecting links $p_k$ and $s_k$, where $p_k$ is the predecessor of joint $k$ and $s_k$ the successor. The velocity of link $s_k$ can be expressed as

$$\hat{\mathbf{v}}_{s_k} = \hat{\mathbf{v}}_{p_k} + \hat{\mathbf{S}}_k \dot{\mathbf{q}}_k , \qquad (9.1)$$

where $\hat{\mathbf{S}}_k$ is the motion space of joint $k$ and $\mathbf{q}_k$ its velocity variables. This equation effectively defines a constraint on the generalised velocity of the tree, since both $\hat{\mathbf{v}}_{p_k}$ and $\hat{\mathbf{v}}_{s_k}$ are functions of the generalised velocity. It is better to express the constraint as

$$\hat{\mathbf{R}}_k{}^S(\hat{\mathbf{v}}_{s_k} - \hat{\mathbf{v}}_{p_k}) = \mathbf{0}\,, \tag{9.2}$$

where $\hat{\mathbf{R}}_k = \hat{\mathbf{S}}_k{}^{\perp}$ is the reaction-force space of joint $k$. The number of constraints imposed on the velocity is equal to the dimension of $\hat{\mathbf{R}}_k$. In subsequent equations, the compound subscript will be abbreviated so that $\hat{\mathbf{v}}_{s_k}$ reads $\hat{\mathbf{v}}_s$, $\hat{\mathbf{v}}_{p_k}$ reads $\hat{\mathbf{v}}_p$, and so on.

Joint $k$ also exerts forces on links $p_k$ and $s_k$. If $\hat{\mathbf{f}}_k$ is the force exerted on link $s_k$ then

$$\hat{\mathbf{f}}_k = \hat{\mathbf{f}}_k^a + \hat{\mathbf{R}}_k\,\mathbf{f}_k\,, \tag{9.3}$$

where $\hat{\mathbf{f}}_k^a$ is the actuation force, $\hat{\mathbf{R}}_k\,\mathbf{f}_k$ is the reaction force, and $\mathbf{f}_k$ is the vector of unknown reaction force coefficients. The number of unknowns is equal to the dimension of $\hat{\mathbf{R}}_k$, and is the same as the number of motion constraints introduced by the loop. The force exerted on link $p_k$ is $-\hat{\mathbf{f}}_k$.

If there is more than one loop joint in the mechanism, then each one imposes its own motion constraints and introduces its own unknown reaction forces. The total number of constraints is equal to the total number of extra unknowns, and will be denoted by $r$. If the motion constraints are independent in the space of generalised velocities then the mechanism has $n-r$ degrees of motion freedom.

In summary, the presence of kinematic loops affects the dynamics of a tree-structure mechanism by introducing a number of loop-closure forces acting between certain links. These forces contain unknown components, but their effect is to impose known motion constraints on the tree.

## 9.3. The Equations of Motion for Robots with Kinematic Loops

The equations of motion of a robot mechanism containing kinematic loops are the equations of motion of a tree-structure mechanism subject to a number of loop-closure forces. The equations of motion of a tree-structure mechanism have the form

$$\mathbf{H}\,\ddot{\mathbf{q}} = \mathbf{Q} - \mathbf{C}\,, \tag{9.4}$$

where $\mathbf{H}$ is the joint-space inertia matrix, $\mathbf{C}$ is the vector of generalised forces accounting for velocity-product effects, gravity, etc., and $\mathbf{Q}$ and $\ddot{\mathbf{q}}$ are vectors of generalised forces and accelerations respectively. Methods for calculating $\mathbf{H}$ and $\mathbf{C}$ are discussed in Chapter 5. Equation (9.4) is a set of $n$ simultaneous linear equations in $n$ unknowns. If we add the generalised loop-closure forces to the right-hand side of this equation, making it the equation of motion for a mechanism containing kinematic loops, then we have introduced another $r$ unknowns. So, in order to solve for the motion of the system, we must supplement the equations of motion with the $r$ motion constraints imposed by the loops.

The procedure for obtaining the equations of motion for a mechanism containing kinematic loops is as follows:

1. for each loop obtain a constraint on the acceleration of the tree mechanism, expressed in terms of its generalised acceleration;

2. express the loop-closure forces in terms of the equivalent generalised forces; and

3. add the loop-closure forces to the equation of motion for the tree, giving the equation of motion for the complete mechanism, and combine this with the acceleration constraint equations.

The result is a system of $n+r$ linear equations in $n+r$ unknowns, where $n$ is the number of degrees of freedom of the tree and $r$ is the total number of loop constraints.

**Step 1: Acceleration constraints**

Loop joint $k$ imposes a constraint on the relative velocity of links $p_k$ and $s_k$ given by Equation (9.2). The constraint on their accelerations can be obtained by differentiating this equation, giving

$$\hat{\mathbf{R}}_k^S(\hat{\mathbf{a}}_s - \hat{\mathbf{a}}_p) + \dot{\hat{\mathbf{R}}}_k^S(\hat{\mathbf{v}}_s - \hat{\mathbf{v}}_p) = \mathbf{0}\,. \tag{9.5}$$

$\hat{\mathbf{a}}_p$ and $\hat{\mathbf{a}}_s$ are the accelerations of links $p_k$ and $s_k$, and $\dot{\hat{\mathbf{R}}}_k$ is the absolute derivative of $\hat{\mathbf{R}}_k$. Equation (9.5) gives us a constraint on the spatial accelerations of two links. We must translate this into a constraint on the generalised acceleration of the tree, $\ddot{\mathbf{q}}$.

The velocity of any link $i$ is given in terms of the velocities of the tree joints by the equation

$$\hat{\mathbf{v}}_i = \sum_{j=1}^{N} e_{ij}\,\hat{\mathbf{S}}_j\,\dot{\mathbf{q}}_j\,, \tag{9.6}$$

where $\hat{\mathbf{S}}_j$ is the motion space of tree joint $j$, $\dot{\mathbf{q}}_j$ is its velocity, and $e_{ij}$ is a scalar taking the value 1 if joint $j$ is on the path between link $i$ and the base and 0 otherwise. The $e_{ij}$ are constants which depend on the connectivity of the tree. This equation can be expressed as a matrix product

$$\hat{\mathbf{v}}_i = \hat{\mathbf{J}}_i\,\dot{\mathbf{q}}\,, \tag{9.7}$$

where $\dot{\mathbf{q}}$ is the generalised velocity of the tree and $\hat{\mathbf{J}}_i$ is the $6 \times n$ matrix

$$\hat{\mathbf{J}}_i = [\, e_{i1}\hat{\mathbf{S}}_1 \;\cdots\; e_{iN}\hat{\mathbf{S}}_N\,]\,. \tag{9.8}$$

$\hat{\mathbf{J}}_i$ is the Jacobian of link $i$, and is a generalisation of the familiar Jacobian matrix which gives the velocity of a robot's end effector in terms of its joint velocities. Differentiating Equation (9.7) gives an expression for the spatial acceleration:

$$\hat{\mathbf{a}}_i = \hat{\mathbf{J}}_i\,\ddot{\mathbf{q}} + \dot{\hat{\mathbf{J}}}_i\,\dot{\mathbf{q}}\,. \tag{9.9}$$

The second term is the acceleration of link $i$ due to velocity-product effects, i.e., the acceleration that link $i$ would have if $\ddot{\mathbf{q}} = \mathbf{0}$. Now this quantity must be calculated for every link in the mechanism in order to compute the velocity-product effects in the vector $\mathbf{C}$ (Equation (9.4)). There is no point in calculating it twice, so let us call the velocity-product acceleration of link $i$ $\hat{\mathbf{a}}_i^{vp}$, and assume that its value is already known as a by-product of calculating $\mathbf{C}$. The acceleration of link $i$ is then given by

$$\hat{\mathbf{a}}_i = \hat{\mathbf{J}}_i \ddot{\mathbf{q}} + \hat{\mathbf{a}}_i^{vp} . \tag{9.10}$$

The acceleration constraint equation, Equation (9.5), can now be expressed in terms of $\ddot{\mathbf{q}}$ as follows:

$$\hat{\mathbf{R}}_k^S ( \hat{\mathbf{J}}_s - \hat{\mathbf{J}}_p ) \ddot{\mathbf{q}} = - \hat{\mathbf{R}}_k^S ( \hat{\mathbf{a}}_s^{vp} - \hat{\mathbf{a}}_p^{vp} ) - \dot{\hat{\mathbf{R}}}_k^S ( \hat{\mathbf{v}}_s - \hat{\mathbf{v}}_p ) . \tag{9.11}$$

Let us introduce the quantities $\hat{\mathbf{v}}_k = \hat{\mathbf{v}}_s - \hat{\mathbf{v}}_p$, $\hat{\mathbf{a}}_k^{vp} = \hat{\mathbf{a}}_s^{vp} - \hat{\mathbf{a}}_p^{vp}$ and $\hat{\mathbf{J}}_k = \hat{\mathbf{J}}_s - \hat{\mathbf{J}}_p$. These are, respectively, the velocity across joint $k$, the velocity-product acceleration across joint $k$, and the Jacobian of loop $k$. Equation (9.11) then simplifies to

$$\hat{\mathbf{R}}_k^S \hat{\mathbf{J}}_k \ddot{\mathbf{q}} = - \hat{\mathbf{R}}_k^S \hat{\mathbf{a}}_k^{vp} - \dot{\hat{\mathbf{R}}}_k^S \hat{\mathbf{v}}_k . \tag{9.12}$$

Equation (9.12) expresses the constraints imposed on the acceleration of the tree by loop $k$. The number of constraints is equal to the dimension of $\hat{\mathbf{R}}_k$, and each loop gives rise to one such constraint equation.

**Step 2: Loop-closure forces**

The loop-closure force for loop $k$ consists of a force $\hat{\mathbf{f}}_k$ exerted on link $s_k$ and $-\hat{\mathbf{f}}_k$ on link $p_k$. We must express these forces in terms of their equivalent generalised forces. Suppose that a generalised force $\mathbf{Q}$ applied to the tree has the effect of statically balancing a force $\hat{\mathbf{f}}$ applied to link $i$ (ignoring all dynamic effects). For this to be so, a force of $-\hat{\mathbf{f}}$ must be transmitted to link $i$ from its predecessor, which in turn receives a force of

$-\hat{\mathbf{f}}$ from its predecessor, and so on. So each joint on the path from the base to link $i$ transmits a force of $-\hat{\mathbf{f}}$ on to the next link. If $j$ is such a joint, then the generalised force for this joint is given by

$$\mathbf{Q}_j = -\,\hat{\mathbf{S}}_j{}^S\,\hat{\mathbf{f}}\,.$$

The generalised force for the whole tree is given by

$$\mathbf{Q} = -\,[\,e_{i1}\hat{\mathbf{S}}_1\ \cdots\ e_{iN}\hat{\mathbf{S}}_N\,]^S\,\hat{\mathbf{f}}\,, \tag{9.13}$$

which, on comparing with the definition of the Jacobian in Equation (9.8), is seen to be

$$\mathbf{Q} = -\,\hat{\mathbf{J}}_i{}^S\,\hat{\mathbf{f}}\,. \tag{9.14}$$

Since this force statically balances $\hat{\mathbf{f}}$ applied to link $i$, it is equivalent to the force $-\hat{\mathbf{f}}$ on link $i$.

If $\mathbf{Q}_k^L$ is the joint force equivalent to the loop-closure forces $\hat{\mathbf{f}}_k$ on $s_k$ and $-\hat{\mathbf{f}}_k$ on $p_k$, then $\mathbf{Q}_k^L$ is given by

$$\begin{aligned}\mathbf{Q}_k^L &= \hat{\mathbf{J}}_s{}^S\,\hat{\mathbf{f}}_k - \hat{\mathbf{J}}_p{}^S\,\hat{\mathbf{f}}_k \\ &= \hat{\mathbf{J}}_k{}^S\,\hat{\mathbf{f}}_k\,.\end{aligned} \tag{9.15}$$

The effect on the dynamics of the tree of closing loop $k$ is the same as the effect of adding $\mathbf{Q}_k^L$ to the generalised force. The effect of closing all loops simultaneously is the same as adding $\sum_{k=1}^{L}\ \mathbf{Q}_k^L$ to the generalised force.

**Step 3: Equations of motion**

The equation of motion of the tree system, taking into account the loop-closure forces, is a system of linear simultaneous equations

$$\mathbf{H}\,\ddot{\mathbf{q}} = \mathbf{Q} - \mathbf{C} + \sum_{k=1}^{L}\ \mathbf{Q}_k^L\,. \tag{9.16}$$

All the forces in the original mechanism are accounted for, so these are effectively the equations of motion for the original mechanism. However, there are more unknowns than equations in this system, so we must augment it with the loop acceleration constraints (Equation (9.12)). The result of combining Equation (9.16) with the acceleration constraints and expanding $\mathbf{Q}_k^L$ in terms of the unknown force coefficients (using Equations (9.15 and 9.3)) is

$$\left[\begin{array}{c|cccc} \mathbf{H} & \hat{\mathbf{J}}_1{}^S\hat{\mathbf{R}}_1 & \hat{\mathbf{J}}_2{}^S\hat{\mathbf{R}}_2 & \cdots & \hat{\mathbf{J}}_L{}^S\hat{\mathbf{R}}_L \\ \hline \hat{\mathbf{R}}_1{}^S\hat{\mathbf{J}}_1 & & & & \\ \hat{\mathbf{R}}_2{}^S\hat{\mathbf{J}}_2 & & & & \\ \vdots & & O & & \\ \hat{\mathbf{R}}_L{}^S\hat{\mathbf{J}}_L & & & & \end{array}\right] \left[\begin{array}{c} \ddot{\mathbf{q}} \\ \hline -\mathbf{f}_1 \\ -\mathbf{f}_2 \\ \vdots \\ -\mathbf{f}_L \end{array}\right] = \left[\begin{array}{c} \mathbf{Q}-\mathbf{C}+\sum_{k=1}^{L}\hat{\mathbf{J}}_k^S\hat{\mathbf{f}}_k^a \\ \hline -\hat{\mathbf{R}}_1^S\hat{\mathbf{a}}_1^{vp}-\hat{\mathbf{R}}_1^S\hat{\mathbf{v}}_1 \\ -\hat{\mathbf{R}}_2^S\hat{\mathbf{a}}_2^{vp}-\hat{\mathbf{R}}_2^S\hat{\mathbf{v}}_2 \\ \vdots \\ -\hat{\mathbf{R}}_L^S\hat{\mathbf{a}}_L^{vp}-\hat{\mathbf{R}}_L^S\hat{\mathbf{v}}_L \end{array}\right] \qquad (9.17)$$

The coefficient matrix is symmetric and has dimension $n+r$, where $n$ is the number of degrees of freedom of the tree and $r$ the total number of constraint forces. The $n \times n$ sub-matrix $\mathbf{H}$ is positive definite, but the matrix as a whole is not, and it is possible for it to be singular (see below). Do not be deceived by the apparent size of the zero area in the matrix -- if the robot is to have any motion freedom at all, then the number of constraints must be less than the number of open-loop degrees of freedom (in general), so the size of the zero portion is less than that of $\mathbf{H}$.

## 9.4. Calculation Considerations

In order to calculate the motion of a robot mechanism containing kinematic loops, Equation (9.17) must be set up and solved for the acceleration of the open-loop mechanism. The accelerations will be consistent with the loop constraints, except for small numerical errors, and they can be integrated to find the position and velocity of the mechanism at the next instant.

The following terms need to be calculated: $\mathbf{H}$ and $\mathbf{C}$ for the open-loop mechanism, the velocity-product terms $\hat{\mathbf{R}}_k^S \hat{\mathbf{a}}_k^{vp} + \hat{\mathbf{R}}_k^S \hat{\mathbf{v}}_k$, the active loop-joint forces $\sum_{k=1}^{L} \hat{\mathbf{J}}_k^S \hat{\mathbf{f}}_k^a$ and the products $\hat{\mathbf{J}}_k^S \hat{\mathbf{R}}_k$.

The calculation of $\mathbf{H}$ and $\mathbf{C}$ is as described in earlier chapters, except that the procedure which calculates $\mathbf{C}$ should make the values it calculates for the link velocities and velocity-product accelerations accessible for the (later) calculation of velocity-product terms. The calculation of the other terms is described below.

The additional computations may be performed in either link or absolute coordinates. If absolute coordinates are used then the quantities $\hat{\mathbf{R}}_k$, $\hat{\mathbf{R}}_k$ and $\hat{\mathbf{f}}_k^a$ must be transformed to absolute coordinates; but if the calculations are to be performed in link coordinates then several additional transformations are required to calculate the terms involving the loop Jacobians. Which is faster depends on the sizes of the loops (i.e., the number of joints involved in each loop) and the dimensions of the various sub-spaces: large loops and low-dimensional $\hat{\mathbf{R}}_k$ favour absolute coordinates. The absolute-coordinate version has the advantage of being easier to implement.

The amount of calculation needed to set up and solve Equation (9.17) is affected by the choice of which joints in the mechanism are treated as loop joints: although $L$ is fixed by the connectivity, $r$ is not, and the smaller the value of $r$ the less calculation is needed. $r$ can be minimised by choosing as loop joints those which impose the smallest numbers of constraints. Also, by choosing un-powered joints some or all of the $\hat{\mathbf{f}}_k^a$ will disappear.

The calculation of the velocity-product terms $\hat{\mathbf{R}}_k^S \hat{\mathbf{a}}_k^{vp} + \hat{\mathbf{R}}_k^S \hat{\mathbf{v}}_k$ is straightforward. If the computations are to be performed in link coordinates, then velocity-product term $k$ should be calculated in the outer frame of loop joint $k$. Remember that $\dot{\hat{\mathbf{R}}}_k$ is the absolute time derivative of $\hat{\mathbf{R}}_k$, not just the component-wise derivative, so it includes the term $\hat{\mathbf{v}}_s \hat{\times} \hat{\mathbf{R}}_k$. Unlike the joint motion space, $\hat{\mathbf{S}}$, which represents both a sub-space and a basis, $\hat{\mathbf{R}}$ represents only a sub-space, so it is sometimes possible to give $\hat{\mathbf{R}}$ a simpler form than $\hat{\mathbf{S}}$. For example, the reaction-force space for a spherical joint is

$$\hat{R} = \begin{bmatrix} 1 & 0 & 0 \\ 0 & 1 & 0 \\ 0 & 0 & 1 \\ 0 & 0 & 0 \\ 0 & 0 & 0 \\ 0 & 0 & 0 \end{bmatrix}$$

regardless of the choice of joint velocity variables.

Calculations involving the loop Jacobians can be implemented using the $L \times N$ matrix of coefficients $E_{ki} = e_{si} - e_{pi}$, which define the circuit of each kinematic loop.[18] The elements of this matrix are constants having the values $+1$, $-1$ or $0$, depending on whether tree joint $i$ is on the path leading to link $s_k$, the path leading to link $p_k$, or is not a part of the circuit of loop $k$. In terms of these coefficients, loop Jacobian $\hat{\mathbf{J}}_k$ is expressed as

$$\hat{\mathbf{J}}_k = [\, E_{k1}\hat{\mathbf{S}}_1 \;\cdots\; E_{kN}\hat{\mathbf{S}}_N \,],$$

and the product $\hat{\mathbf{R}}_k^S \hat{\mathbf{J}}_k$ as

$$\hat{\mathbf{R}}_k^S \hat{\mathbf{J}}_k = [\, E_{k1}\hat{\mathbf{R}}_k^S \hat{\mathbf{S}}_1 \;\cdots\; E_{kN}\hat{\mathbf{R}}_k^S \hat{\mathbf{S}}_N \,].$$

Each individual sub-matrix $\hat{\mathbf{R}}_k^S \hat{\mathbf{S}}_i$ need only be calculated if $E_{ki}$ is non-zero.

---

[18] This matrix is equivalent to the matrix $U$ in [53].

If the computations are to be performed in link coordinates, then for each sub-matrix $\hat{\mathbf{R}}_k^S \hat{\mathbf{S}}_i$ we have a choice as to whether to compute it in $i$-coordinates or in the outer frame of loop joint $k$. The former involves transforming $\hat{\mathbf{R}}_k$ around the links making up the loop, and the latter involves transforming the relevant $\hat{\mathbf{S}}_i$ to joint $k$'s outer frame. Which is faster depends on the dimensions of the quantities involved and the size of the loop; but whichever way is chosen a large amount of transformation must be done, and the total amount of computation is likely to be more than for the corresponding calculations performed in absolute coordinates.

Once Equation (9.17) has been set up, the next step is to solve for the unknown accelerations and constraint forces. Although it is possible for the matrix in Equation (9.17) to be singular, it is always possible to determine the accelerations. The accelerations comply with the motion constraints imposed by the loops, and can be integrated to get the position and velocity of the mechanism at the next time instant.

Due to numerical errors in the calculation and integration of the accelerations, care must be taken to avoid the possibility that errors will accumulate in the position and velocity of the open-loop mechanism which are not consistent with the loop constraints. One way to prevent such errors from growing is to stabilise the constraints [5], [48], [76]. This process is effectively like putting weak springs and dampers in parallel with the loop-closure constraints, so that small errors in position and velocity induce forces which tend to correct the errors. Another way is to identify $n-r$ independent joint variables and calculate the dependent ones explicitly from the independent ones so that the constraints are satisfied exactly. In this case only the independent variables are integrated.

### 9.4.1. Redundant Constraints

It may happen that the matrix $\hat{\mathbf{J}}_k^S \hat{\mathbf{R}}_k$ does not have full rank, or that when assembled into Equation (9.17) some of the columns are linearly dependent on those of other loops. In the first case it means that some of the loop-closure constraints are linearly dependent on constraints in the tree-mechanism part of the loop; and in the second case it indicates that, although topologically independent, the loops are not algebraically independent. Whichever case pertains, the result is that the coefficient matrix in Equation (9.17) is singular and one or more of the loop-closure forces can not be determined. If the indeterminate forces are removed from the equations, then the remainder can be solved for the rest of the unknowns.

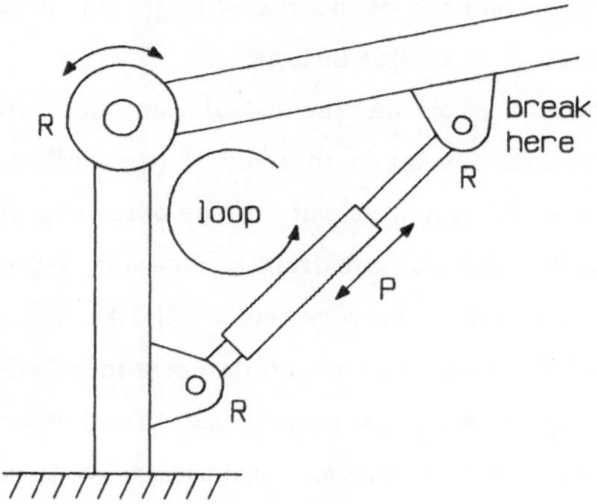


**Figure 9-2:** Piston-actuated joint

This state of affairs can be either transitory (i.e., occurring only at certain positions) or permanent. In the latter case it is usually possible to identify and remove redundant constraints from the mechanism beforehand. For example, consider the piston-actuated revolute joint shown in Figure

9-2. If we cut the loop at the joint shown then the open-loop mechanism has three degrees of motion freedom -- two revolute and one prismatic -- but the loop joint imposes five motion constraints on this mechanism. In fact, three of these constraints are linearly dependent on constraints in the rest of the mechanism -- they serve to prevent non-planar motion when the mechanism is already constrained to have planar motion -- so only two constraints have any effect, and the loop has one degree of motion freedom. The constraint forces for the redundant constraints are indeterminate, and the redundant constraints should be removed from the equations. This can be done by treating the loop joint as a sphere-in-cylinder joint, which imposes the correct number of constraints. A beneficial side-effect is that $r$ is smaller, so less calculation is needed to find the accelerations.

## 9.5. The Articulated-Body Method for Kinematic Loops

The articulated-body method is not easily adapted to cope with kinematic loops: major modifications are required, and the resulting algorithm is highly inefficient. The algorithm described here is not recommended for practical use, but is included mainly for interest and completeness.[19] The algorithm works by calculating the inverse articulated-body and cross inertias for an open-loop mechanism, treating it as an articulated body with many handles. Kinematic loops are modelled as constraint forces acting between pairs of handles, which impose constraints on the relative accelerations of the handles.

Although the algorithm itself is of little practical use, the techniques developed for it may be useful in other respects. For example, knowing the inverse articulated-body inertia of a robot's end effector allows one to calculate directly the effect on the motion of the end effector of forces applied to it by the environment.

[19] A more efficient algorithm is currently being developed by Lathrop at MIT (personal communication). It uses a more general form of constraint propagation.

### 9.5.1. Calculating the Motion of a Mechanism with Kinematic Loops

The equation of motion for an articulated body with $H$ handles is

$$\begin{bmatrix} \hat{\mathbf{a}}_1 \\ \hat{\mathbf{a}}_2 \\ \cdot \\ \cdot \\ \cdot \\ \hat{\mathbf{a}}_H \end{bmatrix} = \begin{bmatrix} \hat{\Phi}_1 & \hat{\Phi}_{12} & \cdots & \hat{\Phi}_{1H} \\ \hat{\Phi}_{21} & \hat{\Phi}_2 & \cdots & \hat{\Phi}_{2H} \\ \cdot & \cdot & \cdots & \cdot \\ \cdot & \cdot & \cdots & \cdot \\ \cdot & \cdot & \cdots & \cdot \\ \hat{\Phi}_{H1} & \hat{\Phi}_{H2} & \cdots & \hat{\Phi}_H \end{bmatrix} \begin{bmatrix} \hat{\mathbf{f}}_1 \\ \hat{\mathbf{f}}_2 \\ \cdot \\ \cdot \\ \cdot \\ \hat{\mathbf{f}}_H \end{bmatrix} + \begin{bmatrix} \hat{\mathbf{b}}_1 \\ \hat{\mathbf{b}}_2 \\ \cdot \\ \cdot \\ \cdot \\ \hat{\mathbf{b}}_H \end{bmatrix}, \tag{9.18}$$

where $\hat{\mathbf{a}}_i$ is the acceleration of handle number $i$, $\hat{\mathbf{b}}_i$ is the bias acceleration of handle $i$, $\hat{\Phi}_i$ is the inverse articulated-body inertia of handle $i$, and $\hat{\Phi}_{ij}$ is the inverse cross-inertia relating a force on handle $j$ to an acceleration on handle $i$. This is a straightforward generalisation of Equation (3.16). Before considering the problem of how to calculate the various inertias, let us first obtain an equation of motion for a mechanism containing kinematic loops.

Let there be $L$ loops, and let handles $p_k$ and $s_k$ be the predecessor and successor links of loop joint $k$. (Handle numbers do not necessarily correspond to link numbers.) Loop joint $k$ imposes an acceleration constraint

$$\hat{\mathbf{R}}_k^S(\hat{\mathbf{a}}_{s_k} - \hat{\mathbf{a}}_{p_k}) + \hat{\mathbf{R}}_k^S(\hat{\mathbf{v}}_{s_k} - \hat{\mathbf{v}}_{p_k}) = \mathbf{0} \tag{9.19}$$

on handles $s_k$ and $p_k$ by applying a force of $\hat{\mathbf{R}}_k\,\mathbf{f}_k$ on $s_k$ and $-\hat{\mathbf{R}}_k\,\mathbf{f}_k$ on $p_k$. Substituting for the accelerations in Equation (9.19) gives

$$\hat{\mathbf{R}}_k^S\Big(\sum_{h=1}^{H}(\hat{\Phi}_{sh}-\hat{\Phi}_{ph})\hat{\mathbf{f}}_h + \hat{\mathbf{b}}_s - \hat{\mathbf{b}}_p\Big) + \hat{\mathbf{R}}_k^S(\hat{\mathbf{v}}_s - \hat{\mathbf{v}}_p) = \mathbf{0}\,, \tag{9.20}$$

where compound subscripts have been abbreviated as before. Let $\beta_k$ be the loop bias acceleration

$$\beta_k = \hat{\mathbf{R}}_k^S(\hat{\mathbf{b}}_s - \hat{\mathbf{b}}_p) + \hat{\mathbf{R}}_k^S(\hat{\mathbf{v}}_s - \hat{\mathbf{v}}_p), \tag{9.21}$$

then Equation (9.20) becomes

$$\sum_{h=1}^{H} \hat{\mathbf{R}}_k^S(\hat{\Phi}_{sh} - \hat{\Phi}_{ph})\hat{\mathbf{f}}_h + \beta_k = \mathbf{0}. \tag{9.22}$$

Now $\hat{\mathbf{f}}_h$ is the sum of the loop-closure forces for all the loops which connect to handle $h$, and is given by

$$\hat{\mathbf{f}}_h = \sum_{l=1}^{L} C_{hl}\hat{\mathbf{R}}_l\mathbf{f}_l, \tag{9.23}$$

where

$$C_{hl} = \begin{cases} 1 & h = s_l \\ -1 & h = p_l \\ 0 & \text{otherwise} \end{cases}. \tag{9.24}$$

Substituting Equation (9.23) into Equation (9.22) gives

$$\sum_{h=1}^{H}\sum_{l=1}^{L} \hat{\mathbf{R}}_k^S(\hat{\Phi}_{sh} - \hat{\Phi}_{ph})C_{hl}\hat{\mathbf{R}}_l\mathbf{f}_l + \beta_k = \mathbf{0}. \tag{9.25}$$

For any given value of $l$, only the values $h = s_l$ and $h = p_l$ contribute to the sum, so Equation (9.25) simplifies to

$$\sum_{l=1}^{L} \hat{\mathbf{R}}_k^S(\hat{\Phi}_{ss} - \hat{\Phi}_{ps} - \hat{\Phi}_{sp} + \hat{\Phi}_{pp})\hat{\mathbf{R}}_l\mathbf{f}_l + \beta_k = \mathbf{0}. \tag{9.26}$$

($\hat{\Phi}_{ss}$ stands for $\hat{\Phi}_{s_k s_l}$, $\hat{\Phi}_{ps}$ for $\hat{\Phi}_{p_k s_l}$, and so on.) Let $A_{kl}$ be the matrix

$$A_{kl} = \hat{\mathbf{R}}_k^S(\hat{\Phi}_{ss} - \hat{\Phi}_{ps} - \hat{\Phi}_{sp} + \hat{\Phi}_{pp})\hat{\mathbf{R}}_l, \tag{9.27}$$

then Equation (9.26) is given by

$$\sum_{l=1}^{L} \Lambda_{kl}\,\mathbf{f}_l + \beta_k = \mathbf{0}\,, \tag{9.28}$$

and combining the equations for all loops gives

$$\begin{bmatrix} \Lambda_{11} & \cdot & \cdot & \cdot & \Lambda_{1L} \\ \cdot & \cdot & \cdot & \cdot & \cdot \\ \cdot & \cdot & \cdot & \cdot & \cdot \\ \cdot & \cdot & \cdot & \cdot & \cdot \\ \Lambda_{L1} & \cdot & \cdot & \cdot & \Lambda_{LL} \end{bmatrix} \begin{bmatrix} \mathbf{f}_1 \\ \cdot \\ \cdot \\ \cdot \\ \mathbf{f}_L \end{bmatrix} + \begin{bmatrix} \beta_1 \\ \cdot \\ \cdot \\ \cdot \\ \beta_L \end{bmatrix} = \mathbf{0}\ . \tag{9.29}$$

Equation (9.29) can be solved for the loop-closure forces (assuming the coefficient matrix is not singular), and once the forces are known they can be substituted into Equation (9.18) to find the accelerations of the handles.

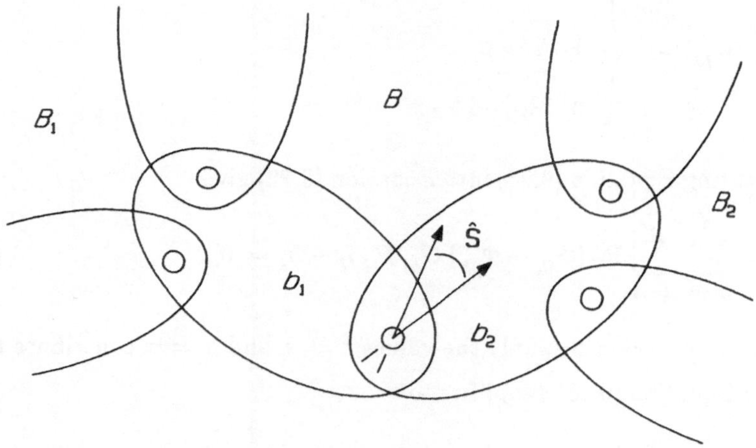


**Figure 9-3:** Part of an articulated body

### 9.5.2. Calculating Several Articulated-Body Inertias from one Articulated Body

Referring to Figure 9-3, let $b_1$ and $b_2$ be the handles of the loop-free articulated bodies $B_1$ and $B_2$ which form the articulated body $B$ when connected as shown. We are interested in calculating the articulated-body inertias of several handles in the one articulated body $B$. Let $b_1$ and $b_2$ have articulated-body inertias $\hat{\mathbf{I}}_1^A$ and $\hat{\mathbf{I}}_2^A$ in $B_1$ and $B_2$ respectively, and $\hat{\mathbf{I}}_1^B$ and $\hat{\mathbf{I}}_2^B$ in $B$. From Equations (6.17 and 7.19) we have that

$$\hat{\mathbf{I}}_1^B = \hat{\mathbf{I}}_1^A + \hat{\mathbf{I}}_2^A - \hat{\mathbf{I}}_2^A\,\hat{\mathbf{S}}(\hat{\mathbf{S}}^S\,\hat{\mathbf{I}}_2^A\,\hat{\mathbf{S}})^{-1}\,\hat{\mathbf{S}}^S\,\hat{\mathbf{I}}_2^A \tag{9.30}$$

and

$$\hat{\mathbf{I}}_2^B = \hat{\mathbf{I}}_1^A + \hat{\mathbf{I}}_2^A - \hat{\mathbf{I}}_1^A\,\hat{\mathbf{S}}(\hat{\mathbf{S}}^S\,\hat{\mathbf{I}}_1^A\,\hat{\mathbf{S}})^{-1}\,\hat{\mathbf{S}}^S\,\hat{\mathbf{I}}_1^A\,, \tag{9.31}$$

where $\hat{\mathbf{S}}$ is the motion sub-space of the joint connecting $b_1$ and $b_2$. Making use of the relation $\hat{\mathbf{I}}_1^B\,\hat{\mathbf{S}} = \hat{\mathbf{I}}_1^A\,\hat{\mathbf{S}}$, $\hat{\mathbf{I}}_2^B$ can be expressed in terms of $\hat{\mathbf{I}}_1^B$ and $\hat{\mathbf{I}}_2^A$ as follows

$$\hat{\mathbf{I}}_2^B = \hat{\mathbf{I}}_1^B + \hat{\mathbf{I}}_2^A\,\hat{\mathbf{S}}(\hat{\mathbf{S}}^S\,\hat{\mathbf{I}}_2^A\,\hat{\mathbf{S}})^{-1}\,\hat{\mathbf{S}}^S\,\hat{\mathbf{I}}_2^A - \hat{\mathbf{I}}_1^B\,\hat{\mathbf{S}}(\hat{\mathbf{S}}^S\,\hat{\mathbf{I}}_1^B\,\hat{\mathbf{S}})^{-1}\,\hat{\mathbf{S}}^S\,\hat{\mathbf{I}}_1^B\,. \tag{9.32}$$

The corresponding equation for bias forces is

$$\begin{aligned}\hat{\mathbf{p}}_2^B = \hat{\mathbf{p}}_1^B &- \hat{\mathbf{I}}_2^A\,\hat{\mathbf{S}}(\hat{\mathbf{S}}^S\,\hat{\mathbf{I}}_2^A\,\hat{\mathbf{S}})^{-1}\,(\mathbf{Q} - \hat{\mathbf{S}}^S\,\hat{\mathbf{p}}_2)\\ &+ \hat{\mathbf{I}}_1^B\,(\,\hat{\mathbf{v}}_2 \hat{\times} \hat{\mathbf{v}}_1 - \hat{\mathbf{S}}(\hat{\mathbf{S}}^S\,\hat{\mathbf{I}}_1^B\,\hat{\mathbf{S}})^{-1}\,\hat{\mathbf{S}}^S\,(\hat{\mathbf{I}}_1^B\,\hat{\mathbf{v}}_2 \hat{\times} \hat{\mathbf{v}}_1 + \hat{\mathbf{p}}_1^B)\,)\,.\end{aligned} \tag{9.33}$$

Equations (9.32 and 9.33) can be converted to recurrence relations for calculating $\hat{\mathbf{I}}_i^B$ and $\hat{\mathbf{p}}_i^B$ for every link in a mechanism by replacing the subscripts 1 and 2 with $\lambda_i$ and $i$ respectively, and fitting a subscript $i$ to $\hat{\mathbf{S}}$ and $\mathbf{Q}$. The calculation proceeds in two stages: first the quantities $\hat{\mathbf{I}}_i^A$ and $\hat{\mathbf{p}}_i$ are calculated using the method described in chapter 6, then the quantities $\hat{\mathbf{I}}_i^B$ and $\hat{\mathbf{p}}_i^B$ are calculated using the recurrence relations based on Equations

(32 and 33). The second stage starts at the base and works out towards the leaves, and the starting values are $\hat{\mathbf{I}}_1^B = \hat{\mathbf{I}}_1^A$ and $\hat{\mathbf{p}}_1^B = \hat{\mathbf{p}}_1$. Making good use of partial results, the second stage involves approximately the same amount of calculation as the first.

### 9.5.3. Calculating Cross-Inertias

The simplest way to calculate cross-inertias is to calculate them from acceleration propagators. The acceleration propagator $\hat{\mathbf{A}}_{ij}$ gives the acceleration of link $i$ (ignoring bias accelerations) in terms of the acceleration of link $j$ which was induced by a test force on link $j$, or on a link connected to link $i$ via link $j$, according to

$$\hat{\mathbf{a}}_i = \hat{\mathbf{A}}_{ij} \hat{\mathbf{a}}_j . \tag{9.34}$$

The inverse cross-inertia is expressed in terms of the acceleration propagator as

$$\hat{\Phi}_{ij} = \hat{\mathbf{A}}_{ij} \hat{\Phi}_j . \tag{9.35}$$

Referring to Figure 9-3, if an acceleration $\hat{\mathbf{a}}_1$ is induced in $b_1$ by a force acting on some link in $B_1$, then the acceleration $\hat{\mathbf{a}}_2$ induced in $b_2$ is given by

$$\hat{\mathbf{a}}_2 = \hat{\mathbf{a}}_1 + \hat{\mathbf{S}} \alpha . \tag{9.36}$$

The force $\hat{\mathbf{f}}$ which produces this acceleration is given by

$$\hat{\mathbf{f}} = \hat{\mathbf{I}}_2^A \hat{\mathbf{a}}_2 \tag{9.37}$$

and satisfies

$$\hat{\mathbf{S}}^S \hat{\mathbf{f}} = \mathbf{0} . \tag{9.38}$$

The acceleration propagator is then calculated as follows:

$$\hat{\mathbf{S}}^S \hat{\mathbf{I}}_2^A ( \hat{\mathbf{a}}_1 + \hat{\mathbf{S}} \alpha ) = \mathbf{0} ,$$

$$\alpha = -(\hat{\mathbf{s}}^S \hat{\mathbf{I}}_2^A \hat{\mathbf{s}})^{-1} \hat{\mathbf{s}}^S \hat{\mathbf{I}}_2^A \hat{\mathbf{a}}_1 ,$$

$$\hat{\mathbf{a}}_2 = (\hat{\mathbf{1}} - \hat{\mathbf{s}}(\hat{\mathbf{s}}^S \hat{\mathbf{I}}_2^A \hat{\mathbf{s}})^{-1} \hat{\mathbf{s}}^S \hat{\mathbf{I}}_2^A) \hat{\mathbf{a}}_1 ,$$

so

$$\hat{\mathbf{A}}_{21} = \hat{\mathbf{1}} - \hat{\mathbf{s}}(\hat{\mathbf{s}}^S \hat{\mathbf{I}}_2^A \hat{\mathbf{s}})^{-1} \hat{\mathbf{s}}^S \hat{\mathbf{I}}_2^A . \tag{9.39}$$

$\hat{\mathbf{I}}_2^B$ may be substituted for $\hat{\mathbf{I}}_2^A$ in Equation (9.39) since $\hat{\mathbf{I}}_2^B \hat{\mathbf{s}} = \hat{\mathbf{I}}_2^A \hat{\mathbf{s}}$. Acceleration propagators across several joints can be constructed as products of acceleration propagators across single joints.

## 9.6. Contact

A contact is a kinematic constraint which arises from physical contact between two rigid bodies which are free to separate. Contacts are useful for describing the motion constraints on objects in the robot's environment. For example, an object resting on the work surface is in contact with the work surface, and is prevented from falling through by the contact forces acting on it; but there is nothing to prevent the robot from picking it up, thereby breaking the contact. If the contact were modelled as a joint then the robot would be able to slide the object over the work surface, but would not be able to pick it up.

The state of contact changes as bodies move around: contacts are made as bodies come together and are broken as they move apart. A motion simulator capable of handling contacts must be prepared to detect the making of new contacts (collision detection) and the breaking of existing ones due to geometric as well as dynamic effects. The geometrical problems will not be considered here; some relevant material can be found in [12], [8], [14], [65], [63]. We will be concerned only with the problem of instantaneous contact dynamics; that is, determining the accelerations given the positions and velocities of the bodies, the instantaneous contact constraints, and the forces acting on the system.

Contact dynamics is a difficult problem to solve in general, so we shall make a number of simplifying assumptions:

- the state of contact between two bodies is described by a finite number of point contacts (Figure 9-4),
- there is no Coulombic friction between contacting bodies, and
- the contact normals are linearly independent in the space of system generalised forces.

The first assumption holds if all contacting bodies are polyhedra, but objects with curved surfaces can cause problems. The third assumption is restrictive, but is only needed by the root-finding algorithm described in Section 9.8.1 for solving the equations of motion. The equations themselves can be formulated independently of this assumption, and there are other methods for solving them which do not require the contact normals to be linearly independent. The problems raised by linearly dependent contact normals will be discussed in Section 9.8.2.

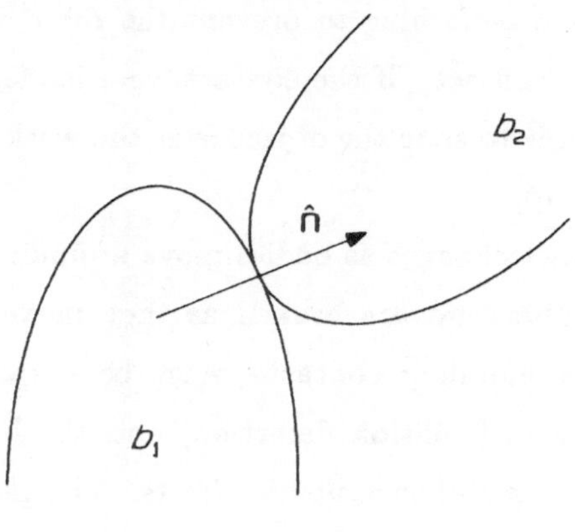


**Figure 9-4:** Point contact

### 9.6.1. Contact Kinematics

A point contact occurs when two bodies meet at a single point.[20] One degree of motion constraint is imposed, which can be modelled by a contact normal vector. Figure 9-4 shows a point contact between bodies $b_1$ and $b_2$. The contact is characterised by a contact normal, $\hat{\mathbf{n}}$, which is a unit line vector (force-type) passing through the point of contact and normal to the surfaces at that point. $\hat{\mathbf{n}}$ is directed from $b_1$ to $b_2$, and the contact force exerted by $b_1$ on $b_2$ is $\alpha\,\hat{\mathbf{n}}$, where $\alpha \geq 0$. ($b_2$ plays the role of the successor body and $b_1$ the predecessor.) If bodies $b_1$ and $b_2$ have velocities of $\hat{\mathbf{v}}_1$ and $\hat{\mathbf{v}}_2$ respectively, then the velocity constraint is [42]

$$\hat{\mathbf{n}}^S(\hat{\mathbf{v}}_2 - \hat{\mathbf{v}}_1) \geq 0 .$$

From the point of view of the instantaneous dynamics, if $\hat{\mathbf{n}}^S(\hat{\mathbf{v}}_2 - \hat{\mathbf{v}}_1) > 0$ then the bodies are already moving apart and the contact can be regarded as broken.

The interesting question is whether or not the two bodies are accelerating away from each other. If they are, then the contact is in the process of breaking, and will be broken immediately after the instant in question. Given that $\hat{\mathbf{n}}^S(\hat{\mathbf{v}}_2 - \hat{\mathbf{v}}_1) = 0$, the acceleration constraint is

$$\hat{\mathbf{n}}^S(\hat{\mathbf{a}}_2 - \hat{\mathbf{a}}_1) + \dot{\hat{\mathbf{n}}}^S(\hat{\mathbf{v}}_2 - \hat{\mathbf{v}}_1) \geq 0 .$$

$\dot{\hat{\mathbf{n}}}$ is the derivative of the contact normal, and is calculated from the velocities of the bodies and the geometry of the contacting surfaces.

For the purposes of instantaneous contact dynamics, the contact is described by the following constraints:

$$\text{either} \quad \hat{\mathbf{n}}^S(\hat{\mathbf{a}}_2 - \hat{\mathbf{a}}_1) + \dot{\hat{\mathbf{n}}}^S(\hat{\mathbf{v}}_2 - \hat{\mathbf{v}}_1) = 0 \quad \text{and} \quad \alpha \geq 0 ,$$

[20] Strictly speaking, we should also require that the local surface curvatures of the bodies be different. See Ohwovoriole [42].

$$\text{or} \qquad \hat{n}^S(\hat{a}_2 - \hat{a}_1) + \dot{\hat{n}}^S(\hat{v}_2 - \hat{v}_1) > 0 \quad \text{and} \quad \alpha = 0\,.$$

In the first case the contact remains and in the second it is breaking. The boundary condition, where both $\alpha$ and the acceleration expression are zero, is called inactive contact. The main difficulty with contact dynamics is that we do not know in advance which case holds for any particular contact: the constraints depend on the forces and the forces on the constraints.

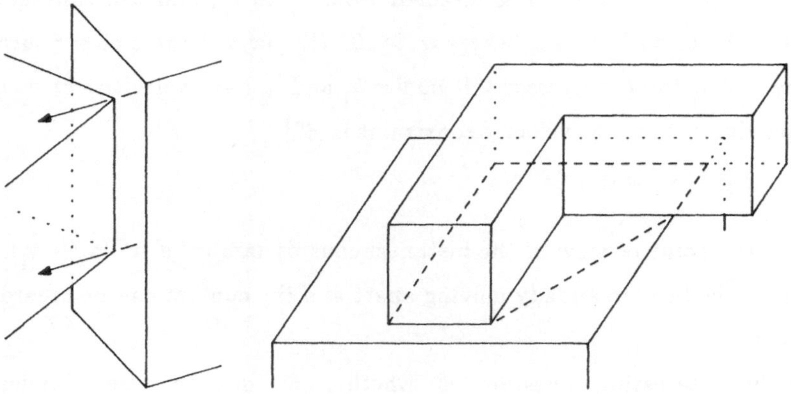

**Figure 9-5:** Extended contacts

Extended contact over curved segments and surface patches must be modelled as a finite number of point contacts. This is possible only for a few special cases. For example, contact along a straight edge segment is dynamically equivalent to two point contacts, one at each end (see Figure 9-5); and contact over a polygonal planar patch can be modelled as a set of point contacts at the vertices of the enclosing convex polygon. On the other hand, contact over a planar circular patch has to be modelled as an infinite number of point contacts, one for each point on the perimeter. In this case the contact normals can only be represented parametrically. More generally, we can say that contact over some region can be modelled by an infinite number of point contacts, one for each point in the region, but that any contact whose contact normal is positively dependent on other contact normals in the region is redundant and can be removed. This means that we can usually disregard most or all of the interior points in a region.

## 9.7. The Equations of Contact Dynamics

The equations of motion for systems of rigid bodies containing contacts can be developed in a similar manner to those of systems containing kinematic loops. In both cases the equation of motion for the constrained system is that of the constraint-free system subject to unknown constraint forces, and the constraint equations are used to solve for the unknowns. The difference lies in the nature of the constraints: loop joints impose equality constraints on the generalised accelerations, whereas contacts impose a mixture of equality and inequality constraints on both the accelerations and the contact forces.

The contact constraint equations can be solved for the unknown contact forces by expressing the generalised accelerations in terms of the contact forces. This results in a system of simultaneous quadratic equations whose roots are subject to a linear inequality. Any root satisfying the inequality is a valid set of contact forces, which can be substituted into the equations of motion of the contact-free system to give the accelerations. The procedure is as follows.

1. Formulate the contact constraint equations in terms of the contact forces and the generalised accelerations of the contact-free system.

2. Use the equations of motion for the contact-free system to express the generalised accelerations in the constraint equations in terms of the contact forces.

An algorithm is described in the next section for finding the roots of the quadratic equations.

**Step 1: Contact constraints**

Let there be $n_c$ contacts, numbered $1 \,.\,.\, n_c$, and let contact $k$ occur between bodies $b_i$ and $b_j$. The contact is characterised by its instantaneous contact normal, $\hat{\mathbf{n}}_k$, and its derivative, $\dot{\hat{\mathbf{n}}}_k$. $\hat{\mathbf{n}}_k$ is directed from $b_i$ to $b_j$, so if $\alpha_k$ is the scalar contact force then the spatial force on $b_j$ is $\hat{\mathbf{n}}_k\alpha_k$ and that on $b_i$ is $-\hat{\mathbf{n}}_k\alpha_k$.

The constraints imposed by the contact on $\alpha_k$ and the relative acceleration of $b_j$ with respect to $b_i$ are either

$$\alpha_k = 0 \quad \text{and} \quad \hat{\mathbf{n}}_k^S \hat{\mathbf{a}}_k + \dot{\hat{\mathbf{n}}}_k^S \hat{\mathbf{v}}_k > 0$$

if the contact is breaking, or

$$\alpha_k \geq 0 \quad \text{and} \quad \hat{\mathbf{n}}_k^S \hat{\mathbf{a}}_k + \dot{\hat{\mathbf{n}}}_k^S \hat{\mathbf{v}}_k = 0$$

if the contact remains. $\hat{\mathbf{v}}_k$ is the relative velocity of $b_j$ with respect to $b_i$ (i.e., $\hat{\mathbf{v}}_k = \hat{\mathbf{v}}_j - \hat{\mathbf{v}}_i$), and $\hat{\mathbf{a}}_k$ the relative acceleration. The constraints are better reformulated as

$$\alpha_k \,( \hat{\mathbf{n}}_k^S \hat{\mathbf{a}}_k + \dot{\hat{\mathbf{n}}}_k^S \hat{\mathbf{v}}_k ) = 0 \tag{9.40}$$

and

$$\alpha_k + \hat{\mathbf{n}}_k^S \hat{\mathbf{a}}_k + \dot{\hat{\mathbf{n}}}_k^S \hat{\mathbf{v}}_k \geq 0\,, \tag{9.41}$$

which apply regardless of whether or not the contact is breaking. Notice that Equation (9.41) combines quantities with different physical dimensions (force and work). This is because it is an abbreviation of two separate inequalities, $\alpha_k \geq 0$ and $\hat{\mathbf{n}}_k^S \hat{\mathbf{a}}_k + \dot{\hat{\mathbf{n}}}_k^S \hat{\mathbf{v}}_k \geq 0$, which is made possible by Equation (9.40).

$\hat{\mathbf{a}}_k$ can be expressed in terms of the generalised acceleration, $\ddot{\mathbf{q}}$, in the same way as was done for loop joints:

$$\hat{\mathbf{a}}_k = \hat{\mathbf{J}}_k \ddot{\mathbf{q}} + \hat{\mathbf{a}}_k^{vp}\,. \tag{9.42}$$

$\hat{\mathbf{J}}_k$ is the relative Jacobian between $b_j$ and $b_i$, and $\hat{\mathbf{a}}_k^{vp}$ is the relative velocity-product acceleration. Similarly, the spatial force $\hat{\mathbf{n}}_k \alpha_k$ between $b_j$ and $b_i$ is equivalent to the generalised force $\mathbf{f}_k \alpha_k$ acting on the system, where

$$\mathbf{f}_k = \hat{\mathbf{J}}_k{}^S \hat{\mathbf{n}}_k \,. \tag{9.43}$$

Using Equations (9.42 and 9.43), the constraint equations for contact $k$ can be expressed in terms of the generalised acceleration as

$$\alpha_k \, ( \, \mathbf{f}_k{}^T \ddot{\mathbf{q}} + \hat{\mathbf{n}}_k{}^S \hat{\mathbf{a}}_k^{vp} + \dot{\hat{\mathbf{n}}}_k{}^S \hat{\mathbf{v}}_k \, ) = 0 \tag{9.44}$$

and

$$\alpha_k + \mathbf{f}_k{}^T \ddot{\mathbf{q}} + \hat{\mathbf{n}}_k{}^S \hat{\mathbf{a}}_k^{vp} + \dot{\hat{\mathbf{n}}}_k{}^S \hat{\mathbf{v}}_k \, \geq \, 0 \,. \tag{9.45}$$

The combined constraint equations for all $n_c$ contacts are given by

$$\mathrm{diag}(\alpha) \, ( \, \mathbf{F}^T \ddot{\mathbf{q}} + \mathbf{u} \, ) = \mathbf{0} \tag{9.46}$$

and

$$\alpha + \mathbf{F}^T \ddot{\mathbf{q}} + \mathbf{u} \, \geq \, \mathbf{0} \,, \tag{9.47}$$

where $\mathbf{F}$ is the $n \times n_c$ matrix composed of the vectors $\mathbf{f}_k$, $\mathbf{u}$ is an $n_c \times 1$ vector whose elements are given by

$$u_k = \hat{\mathbf{n}}_k{}^S \hat{\mathbf{a}}_k^{vp} + \dot{\hat{\mathbf{n}}}_k{}^S \hat{\mathbf{v}}_k \,,$$

$\alpha$ is the $n_c \times 1$ vector of unknown contact forces, and $\mathrm{diag}(\alpha)$ is an $n_c \times n_c$ diagonal matrix containing the elements of $\alpha$.

**Step 2: Eliminating the accelerations from the constraint equations**

The equation of motion for a contact-free system of rigid bodies can be expressed as

$$\mathbf{I} \, \ddot{\mathbf{q}} = \tau \,, \tag{9.48}$$

where $\mathbf{I}$ is a matrix of inertia coefficients, $\tau$ is a vector of generalised forces due to gravity, actuator forces, etc., and $\ddot{\mathbf{q}}$ is the vector of generalised accelerations. If there are no kinematic loops in the system then $\mathbf{I}$ and $\tau$ can be obtained from Equation (5.6), and $\ddot{\mathbf{q}}$ is the vector of joint accelerations. If there are kinematic loops then $\ddot{\mathbf{q}}$ is a vector of independent joint accelerations, and $\mathbf{I}$ and $\tau$ are obtained by eliminating the dependent accelerations and loop-closure forces from Equation (9.17). In both cases $\mathbf{I}$ is symmetric and positive definite.

The effect of introducing contacts into the system is to add a term involving the contact forces to the equation of motion:

$$\begin{aligned}\mathbf{I}\,\ddot{\mathbf{q}} &= \tau + \sum_{k=1}^{n_c} \mathbf{f}_k\, \alpha_k \\ &= \tau + \mathbf{F}\,\alpha\,. \end{aligned} \tag{9.49}$$

Equation (9.49) gives us an expression for $\ddot{\mathbf{q}}$ in terms of $\alpha$, which we can use to substitute for the accelerations in Equations (9.46 and 9.47) resulting in the following equations:

$$\mathrm{diag}(\,\alpha\,)\,(\,\mathbf{C}\,\alpha + \mathbf{d}\,) = \mathbf{0} \tag{9.50}$$

and

$$(\,\mathbf{C} + \mathbf{1}\,)\,\alpha + \mathbf{d} \;\geq\; \mathbf{0}\,, \tag{9.51}$$

where

$$\mathbf{C} = \mathbf{F}^T \mathbf{I}^{-1}\,\mathbf{F} \tag{9.52}$$

and

$$\mathbf{d} = \mathbf{F}^T \mathbf{I}^{-1}\,\tau + \mathbf{u}\,. \tag{9.53}$$

$\mathbf{I}$ is symmetric and positive definite, so $\mathbf{I}^{-1}$ always exists and $\mathbf{C}$ is symmetric and either positive definite or positive semi-definite. The latter case applies if the set of contact normals is linearly dependent in the space of system generalised forces.

Equation (9.50) is a set of $n_c$ simultaneous quadratic equations having, in general, $2^{n_c}$ roots. Any one of these roots which also satisfies the inequality in Equation (9.51) is a valid set of contact forces, and can be substituted into Equation (9.49) to find the acceleration of the system. All valid roots give the same value for the acceleration.

## 9.8. Calculating Contact Dynamics

The motion of a rigid-body system containing contacts is found by calculating $\mathbf{I}$ and $\tau$ for the contact-free system, calculating $\mathbf{C}$ and $\mathbf{d}$ from Equations (9.52 and 9.53), finding a root of Equation (9.50) which satisfies Equation (9.51), and substituting the value of $\alpha$ into Equation (9.49) to find the accelerations. The main problem is to solve Equation (9.50) subject to Equation (9.51).

Equation (9.50) is a set of $n_c$ quadratic equations, of which the $i^{th}$ equation reads

$$\alpha_i \, ( \mathbf{C}_i \, \alpha + d_i ) = 0 \,, \tag{9.54}$$

where $\mathbf{C}_i$ is row $i$ of $\mathbf{C}$. This equation is quadratic in $\alpha_i$ and linear in the other elements of $\alpha$. There are two roots for $\alpha_i$:

$$\alpha_i = 0 \tag{9.55}$$

and

$$\alpha_i = \frac{-1}{C_{ii}} ( \mathbf{C}_{i\hat{i}} \, \alpha_{\hat{i}} + d_i ) \,. \tag{9.56}$$

The notation $\hat{i}$ in subscripts means 'all except $i$'; so $\mathbf{C}_{i\hat{i}}$ is row $i$ of $\mathbf{C}$ with element $C_{ii}$ removed, and $\alpha_{\hat{i}}$ is $\alpha$ with element $\alpha_i$ removed. For simplicity, the roots given by Equations (9.55 and 9.56) will be referred to as the zero and non-zero roots respectively, although it is of course possible for the latter to be zero, making $\alpha_i = 0$ a double root. Physically, a zero root corresponds to a breaking contact and a non-zero root to a remaining contact. We may choose independently for each equation whether to adopt the zero root or the non-zero one. Over the set of $n_c$ equations this leads to a total of $2^{n_c}$ patterns of root choices.

Given that $\beta$ is a root of Equation (9.50), it follows that $\beta_{\bar{i}}$ is a root of

$$\mathrm{diag}(\alpha)\,(\,\mathbf{C}'\,\alpha + \mathbf{d}'\,) = \mathbf{0}\,, \tag{9.57}$$

where $\mathbf{C}'$ and $\mathbf{d}'$ are given by

$$\mathbf{C}' = \mathbf{C}_{\bar{i}\bar{i}} \quad \text{and} \quad \mathbf{d}' = \mathbf{d}_{\bar{i}} \tag{9.58}$$

if $\beta_i$ is a zero root, or

$$\mathbf{C}' = \mathbf{C}_{\bar{i}\bar{i}} - \frac{\mathbf{C}_{\bar{i}i}\,\mathbf{C}_{i\bar{i}}}{C_{ii}} \quad \text{and} \quad \mathbf{d}' = \mathbf{d}_{\bar{i}} - \frac{\mathbf{C}_{\bar{i}i}\,d_i}{C_{ii}} \tag{9.59}$$

if $\beta_i$ is a non-zero root. Furthermore, if $\beta$ satisfies Equation (9.51) then $\beta_{\bar{i}}$ satisfies

$$(\,\mathbf{C}' + \mathbf{1}\,)\,\alpha + \mathbf{d}' \,\geq\, \mathbf{0}\,. \tag{9.60}$$

If $\mathbf{C}$ is positive definite then so is $\mathbf{C}'$; but if $\mathbf{C}$ is positive semi-definite then $\mathbf{C}'$ can be either positive definite or semi-definite.

Given a pattern of root choices, it follows from Equations (9.57 and 9.58) that the non-zero roots satisfy the linear equation

$$\mathbf{C}''\,\alpha + \mathbf{d}'' = \mathbf{0}\,, \tag{9.61}$$

where $\mathbf{C}''$ and $\mathbf{d}''$ are obtained from $\mathbf{C}$ and $\mathbf{d}$ by striking out all rows and columns of $\mathbf{C}$ and all elements of $\mathbf{d}$ corresponding to the zero roots. If $\mathbf{C}$ is positive definite then so is $\mathbf{C}''$, and there is a unique solution to Equation (9.61). So each root-choice pattern gives rise to a single system root, and all possible roots are generated by trying every pattern. If $\mathbf{C}$ is positive semi-definite then $\mathbf{C}''$ may be either positive definite or semi-definite. In the latter case there are either an infinite number of solutions to Equation (9.61) or none at all, depending on whether or not it is consistent. (Inconsistency of Equation (9.61) is only possible if C and d form an inconsistent set of linear equations.)

For C″ to be positive semi-definite, the contacts corresponding to the non-zero roots must be linearly dependent; i.e., we are postulating non-zero contact forces on a set of linearly dependent contact normals. If this root choice leads to a root satisfying Equation (9.51) then it is possible to construct another root, also satisfying Equation (9.51), with the property that non-zero contact forces appear only on an independent subset of the contacts. This second root is obtained from the first by adding contact forces which cancel out in the system as a whole. The second root can also be found from a different root-choice pattern -- one that has non-zero roots on a linearly independent set of contacts. So we can conclude that if a solution exists then one exists where the set of non-zero roots corresponds to a set of linearly independent contacts. Effectively we are restricting C″ to be positive definite, and therefore restricting the number of roots to be considered to be less than or equal to $2^{n_c}$.

If Equation (9.50) has a root which satisfies Equation (9.51), then it is possible to find it by trying in turn each pattern of root choices which results in a non-singular C″. The first root found satisfying the constraints is taken as the solution. (The possibility that there is no solution will be discussed later.) The difficulty with this approach is that the number of roots grows exponentially with the number of contacts in the system. A system with 10 contacts, for example, has over a thousand roots, but only a few are likely to satisfy the constraints. The large number of roots is a feature of the problem itself, but we can direct the search for the solution so that we find it more quickly.

### 9.8.1. The Positive Definite Case

If C is positive definite then it can be shown that there is exactly one root (which may be a multiple root) which satisfies the inequality. Since any generated C′ is also positive definite, this also applies to any reduced-dimension system formed from the original problem. That is, if $\beta$ is the only root of Equation (9.50) satisfying Equation (9.51) then $\beta_1^-$ is the only root of Equation (9.57) satisfying Equation (9.60). We can use these facts to formulate a root-finding algorithm which is more efficient than simply trying each pattern in turn.

solve( $n$, $\mathbf{C}$, $\mathbf{d}$, **returning** $\alpha$ )
**begin**
  **if** $n = 1$ **then**
    **if** $d_1 < 0$ **then** $\alpha_1 \leftarrow -\frac{d_1}{C_{11}}$ **else** $\alpha_1 \leftarrow 0$ **endif**
  **else**
    $\alpha_1 \leftarrow 0$;
    solve( $n-1$, $\mathbf{C}_{11}^{--}$, $\mathbf{d}_1^-$, $\alpha_1^-$ );
    **if** $\mathbf{C}_{11}^- \alpha_1^- + d_1 < 0$ **then**
      solve( $n-1$, $\mathbf{C}_{11}^{--} - \frac{\mathbf{C}_{11}^- \mathbf{C}_{11}^-}{C_{11}}$, $\mathbf{d}_1^- - \frac{\mathbf{C}_{11}^- d_1}{C_{11}}$, $\alpha_1^-$ );
      $\alpha_1 \leftarrow \frac{-1}{C_{11}}( \mathbf{C}_{11}^- \alpha_1^- + d_1 )$;
    **endif**
  **endif**
**end**

**Figure 9-6:** Contact dynamics algorithm for positive-definite case

A root-finding algorithm for the positive-definite case is shown in Figure 9-6. The method proceeds as follows. We start by choosing $\alpha_1=0$. This implies that $\alpha_1^-$ is the solution to the reduced-dimension problem obtained by striking out row and column 1 of C and element 1 of d (Equation (9.58)).

We solve the reduced-dimension problem with a recursive call, which returns the only solution to that problem. If $\alpha_1$ and $\alpha_{\bar{1}}$ together satisfy the first constraint equation then $\alpha$ is the overall solution. If they do not, then $\alpha_1=0$ was the wrong choice and we must try again with the non-zero root. This time we use Equation (9.59) to find the reduced-dimension problem, and again solve it recursively. This time there is no need to check the constraint, since we know that our choice for $\alpha_1$ is correct. It remains only to calculate $\alpha_1$ using Equation (9.56). In this way the solution to the original problem is found by solving recursively problems with smaller dimension until a one-dimensional problem is reached, which can be solved directly. It is, of course, perfectly possible to reverse the order of the choices.

The algorithm implements a depth-first search of the root-choice tree. If at any stage in the search the (local) assumption $\alpha_1 = 0$ proves to be correct, then we can prune (i.e., ignore) the branch from $\alpha_1 \neq 0$, which is guaranteed not to contain the solution. To give some idea of the effect of the pruning, if we assume that zero and non-zero roots are equally likely in the overall solution, then this algorithm tries roughly $1.5^{n_c}$ patterns before finding the solution. If $n_c = 10$ then it tries roughly 58 of the 1024 possible patterns.

The performance of the algorithm can be improved further by using a heuristic to guide the choice of zero versus non-zero root at each stage of the search. The heuristic described below has been found to be highly accurate on random-number tests, and on tests with a single rigid body subjected to various conditions of contact and random applied forces. For the random-number tests, C was constructed by multiplying a random matrix with its transpose. Random numbers with a uniform distribution in the range $-1$ to 1 were used. The heuristic was found to be around 98% to 99% accurate on the random-number tests, and from 95% to 100% accurate on the rigid-body tests.

Consider the possibility that $\mathbf{d} \geq \mathbf{0}$. If this is the case then the problem can be solved by inspection, the solution being $\alpha = \mathbf{0}$. The existence of an element of $\mathbf{d}$ which is less than zero implies that at least one element of $\alpha$ must be greater than zero. If there is exactly one negative element in $\mathbf{d}$ then it can be shown that the corresponding element of $\alpha$ is greater than zero, although this is not necessarily the only non-zero element in $\alpha$. If several elements of $\mathbf{d}$ are negative then at least one has a corresponding non-zero element in $\alpha$. With random values for $\mathbf{C}$ and $\mathbf{d}$ there is a strong correlation between the value of an element of $\mathbf{d}$ and the probability that the corresponding element of $\alpha$ will be non-zero -- the more negative the element of $\mathbf{d}$, especially in relation to other elements, the greater the likelihood that the corresponding element of $\alpha$ is non-zero. The correlation gets stronger as the proportion of negative elements in $\mathbf{d}$ goes down.

This suggests that the following heuristic be used to guide the search: find the most negative element of $\mathbf{d}$; if there are no negative elements then return the answer $\alpha = \mathbf{0}$, otherwise assume that the corresponding element of $\alpha$ is positive.

The heuristic is easily incorporated into the pruning algorithm -- all that is required is a search on the elements of $\mathbf{d}$ and the ability to eliminate elements at random (rather than in a fixed order). The effect of the heuristic is that the algorithm picks elements of $\alpha$ which are likely to be non-zero, eliminates them, and is able to tell very quickly whether or not there are any more non-zero $\alpha$'s.

### 9.8.2. The Positive Semi-definite Case

If $\mathbf{C}$ is positive semi-definite then the set of contact normals is linearly dependent in the space of system generalised forces. Unfortunately, we can not ignore contacts with linearly dependent contact normals -- a contact is only redundant if its contact normal is positively dependent on other contact normals. For example, consider the rigid-body system shown in Figure 9-7,

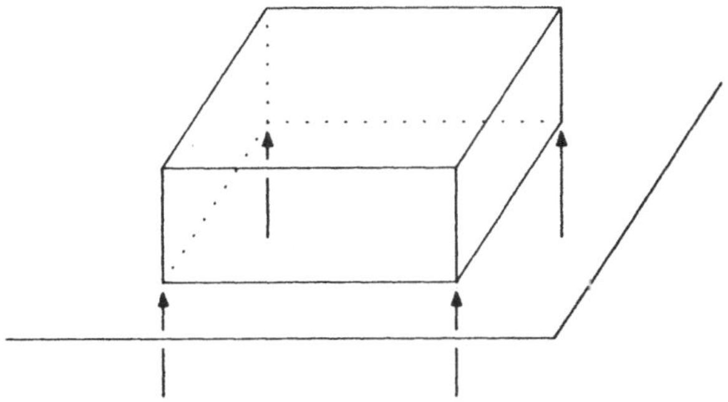

**Figure 9-7:** System with linearly dependent contacts

which consists of a single block resting on a planar support surface. Excluding the contact constraints, the block has six degrees of freedom. The state of contact is described by four point contacts, one at each of the bottom four corners of the block. The contact normals are all parallel, so one is linearly dependent on the other three; but we can not ignore any of the contacts *a priori*. It is not until we know the forces acting on the block that we can decide which three contacts are really necessary for that particular applied force.

From the computational point of view, the main difference between the positive definite and semi-definite cases is that with the latter we are not guaranteed a single solution for the contact forces. There may be zero, one, or infinitely many solutions. This means that the positive semi-definite case cannot be solved using the pruning algorithm of the previous section, since that algorithm relies on the single-solution property of the positive definite case. It is still possible to solve the problem, but we must adopt a more general approach. Equations (9.50 and 9.51) can be treated as a linear complementarity problem, or they can be reformulated into an equivalent quadratic programming problem [37, 38]. The algorithms available for solving these kinds of problems can cope with both the positive definite and semi-definite cases.

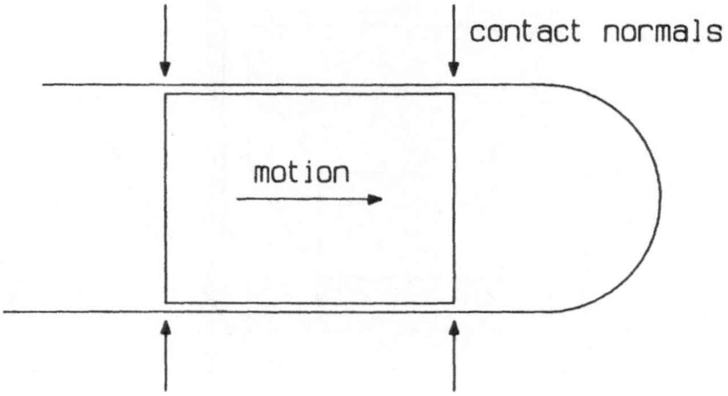


**Figure 9-8:** Second-order collision

The possibility that there may be no solution comes about as the result of a 'second-order' collision: a form of collision brought about by a discontinuity in surface curvature. A 2-dimensional example is shown in Figure 9-8. A rectangular object is moving in a slot which it fits exactly. The end of the slot is curved, and when the object reaches the point at which the curve begins it must stop; but the contact normals are all perpendicular to the direction of motion at this point, and no combination of (finite) forces or impulses can arrest the motion. The linear equations $\mathbf{C}\,\alpha + \mathbf{d} = \mathbf{0}$ suddenly become inconsistent at this point, the inconsistency coming from the term $\mathbf{u}$ in Equation (9.53) which depends on the derivatives of the contact normals. In practice, the block must be allowed to move slightly into the curved region, so that the directions of the contact normals change.

It is sometimes possible to simplify the problem by considering the nature of the dependencies between contact normals. If a set, $\chi$, of contact normals are linearly dependent, then there exist coefficients $a_i$ such that

$$\sum_{i \in \chi} a_i \mathbf{f}_i = \mathbf{0}$$

and not all the $a_i$ are zero. A set is minimal if the removal of any member results in the remainder being linearly independent. The $a_i$ for a minimal set are strictly non-zero. If a minimal set is found with the property that all the coefficients have the same sign, then the contacts in that set can not be broken -- any force in the subspace spanned by the normals is positively dependent on them, and a positive acceleration on one is not possible without an illegal negative acceleration on another. Such contacts can be treated as joints. If a set can be found where one coefficient has one sign and all the rest have the other, then the normal associated with the different sign is positively dependent on the others, and the contact is therefore redundant. As a special case, if a contact normal is the null vector then the contact is redundant.

A substantial amount of computation is involved in analysing the nature of the dependencies, so the analysis is only worthwhile if its results remain valid over a significant period of time (e.g., several integration time steps).

## 9.9. Impact

The equations of contact dynamics take the current set of contacts and work out which, if any, are in the process of breaking due to the bodies accelerating away from each other. Contacts can also be broken by geometric effects, such as a vertex running off the edge of a face, and new contacts are created whenever two bodies, or features of two bodies, collide. In general, at the moment of collision the velocities of the two bodies concerned are inconsistent with the new contact constraints, and unless the velocities are adjusted immediately the two bodies will penetrate each other. Instantaneous velocity changes can be achieved by applying impulses, and the phase of motion during which impulses are acting on the system will be called impact.

An impulse is a mathematical idealisation of a very large force applied over a very short interval: an impulse of magnitude $f'$ is obtained in the limit as $\delta t \rightarrow 0$ of a force of magnitude $f'/\delta t$ applied over an interval $\delta t$. The equations of impulsive motion relate impulses to instantaneous velocity changes in much the same way that the equations of continuous motion relate forces to accelerations. The equations of impulsive motion can be obtained from their continuous-motion counterparts via the limiting process mentioned above.[21] The resulting equations are simpler than their continuous-motion counterparts, since forces which remain finite in the limit, such as gravitational, velocity-product and actuator forces, have no effect on the motion.

The equation of impulsive motion for a rigid-body system with tree structure, whose continuous-motion equation is Equation (5.6), is given by

$$\mathbf{H}\,\delta\dot{\mathbf{q}} = \mathbf{0}\,; \tag{9.62}$$

and that for a system with kinematic loops, whose continuous-motion equation is Equation (9.17), is given by

$$\left[\begin{array}{c|cccc} \mathbf{H} & \hat{\mathbf{J}}_1{}^S\hat{\mathbf{R}}_1 & \hat{\mathbf{J}}_2{}^S\hat{\mathbf{R}}_2 & \cdots & \hat{\mathbf{J}}_L{}^S\hat{\mathbf{R}}_L \\ \hline \hat{\mathbf{R}}_1{}^S\hat{\mathbf{J}}_1 & & & & \\ \hat{\mathbf{R}}_2{}^S\hat{\mathbf{J}}_2 & & & & \\ \vdots & & 0 & & \\ \hat{\mathbf{R}}_L{}^S\hat{\mathbf{J}}_L & & & & \end{array}\right] \left[\begin{array}{c} \delta\dot{\mathbf{q}} \\ \hline -\mathbf{f}'_1 \\ -\mathbf{f}'_2 \\ \vdots \\ -\mathbf{f}'_L \end{array}\right] = \mathbf{0}\,. \tag{9.63}$$

$\delta\dot{\mathbf{q}}$ is the instantaneous velocity change, and $\mathbf{f}'_k$ are the loop-closure impulses. In each case, if an impulse is applied to the system then a vector of generalised impulses appears on the right-hand side.

---

[21] The situation is not quite as simple as this -- several assumptions are involved [76].

### 9.9.1. Impact Between two Rigid Bodies

Suppose that two unconstrained rigid bodies, $b_1$ and $b_2$, collide forming a point contact with contact normal $\hat{\mathbf{n}}$ directed from $b_1$ to $b_2$. An impact occurs, which is broken down into a compression phase and a decompression phase.[22] At the end of the compression phase, the component of the relative velocity along the contact normal is zero; i.e., if the velocities of $b_1$ and $b_2$ before impact are $\hat{\mathbf{v}}_1$ and $\hat{\mathbf{v}}_2$ respectively, and the velocity changes during compression are $\delta\hat{\mathbf{v}}_1^c$ and $\delta\hat{\mathbf{v}}_2^c$, then

$$\hat{\mathbf{n}}^S(\hat{\mathbf{v}}_2 + \delta\hat{\mathbf{v}}_2^c - \hat{\mathbf{v}}_1 - \delta\hat{\mathbf{v}}_1^c) = 0\,. \tag{9.64}$$

Assuming that there is no friction, the impulses applied to $b_1$ and $b_2$ during compression are $-\gamma^c\,\hat{\mathbf{n}}$ and $\gamma^c\,\hat{\mathbf{n}}$ respectively, where $\gamma^c$ is the unknown compression impulse magnitude. To calculate $\gamma^c$ we need to know the inertias of the bodies. If $b_1$ and $b_2$ have inertias $\hat{\mathbf{I}}_1$ and $\hat{\mathbf{I}}_2$ respectively, then

$$\delta\hat{\mathbf{v}}_1^c = -\,\hat{\mathbf{I}}_1^{-1}\,\hat{\mathbf{n}}\,\gamma^c \tag{9.65}$$

and

$$\delta\hat{\mathbf{v}}_2^c = \hat{\mathbf{I}}_2^{-1}\,\hat{\mathbf{n}}\,\gamma^c\,. \tag{9.66}$$

Substituting for $\delta\hat{\mathbf{v}}_1^c$ and $\delta\hat{\mathbf{v}}_2^c$ in Equation (9.64) gives

$$\hat{\mathbf{n}}^S(\hat{\mathbf{v}}_2 - \hat{\mathbf{v}}_1) + \hat{\mathbf{n}}^S(\hat{\mathbf{I}}_1^{-1} + \hat{\mathbf{I}}_2^{-1})\,\hat{\mathbf{n}}\,\gamma^c = 0\,,$$

$$\gamma^c = \frac{-\hat{\mathbf{n}}^S(\hat{\mathbf{v}}_2 - \hat{\mathbf{v}}_1)}{\hat{\mathbf{n}}^S(\hat{\mathbf{I}}_1^{-1} + \hat{\mathbf{I}}_2^{-1})\,\hat{\mathbf{n}}}\,. \tag{9.67}$$

Note that $\gamma^c > 0$ since it is necessary that $\hat{\mathbf{n}}^S(\hat{\mathbf{v}}_2 - \hat{\mathbf{v}}_1) < 0$ for there to have been a collision in the first place.

---

[22] Since we are dealing with rigid bodies, the extent of the compression is zero. The important point is that a certain amount of work is done in compressing the bodies, and some proportion of the energy may be returned to the system during decompression.

During the decompression phase we assume that an impulse $\gamma^d$ is applied, where

$$\gamma^d = e\,\gamma^c\,. \tag{9.68}$$

$e$ is the coefficient of restitution, and may take any value between 0 and 1. The case $e$=0 implies that the collision is perfectly inelastic, and the bodies are in a state of contact after the impact. If $e > 0$ then the bodies bounce apart, and if $e$=1 then the collision is perfectly elastic. The total impulse is the sum of the compression and decompression impulses, and is given by

$$\gamma = -\,(\,1 + e\,)\,\frac{\hat{\mathbf{n}}^S\,(\,\hat{\mathbf{v}}_2 - \hat{\mathbf{v}}_1\,)}{\hat{\mathbf{n}}^S\,(\,\hat{\mathbf{I}}_1^{-1} + \hat{\mathbf{I}}_2^{-1}\,)\,\hat{\mathbf{n}}}\,. \tag{9.69}$$

### 9.9.2. Impact in a Rigid-Body System Containing Contacts

When an impact occurs in a rigid-body system containing contacts, impulses may be generated on several contacts. In general, different contacts will compress at different rates, depending on the properties of the contacting materials and the apparent inertias seen at the contacts. This results in a complex interplay between contacts, which can be described only by using more detailed models of the processes involved (e.g., [9]). For simplicity, we will assume that all contacts compress and decompress synchronously, so that the impact can be divided into distinct compression and decompression phases. It must be borne in mind that this is not generally a valid assumption, so the equations described below may not be adequate for describing the motion of a system during impact.

Let us suppose that a collision has occurred in a rigid-body system containing contacts, creating one or more new contacts. The system velocity at this instant may not comply with all the contact constraints, and must be adjusted by applying impulses at some or all of the contacts. An impact occurs, which is broken down into a compression phase and a decompression phase. During the compression phase, impulses are applied to make the system velocity comply with the contact constraints. During the decompression phase, impulses are applied depending on the compression impulses, the coefficients of restitution and the contact constraints.

Let the system velocity before impact be $\dot{\mathbf{q}}$, and let the change during the compression phase be $\delta\dot{\mathbf{q}}^c$. The contact constraints are, for each contact, either

$$\gamma_k^c \geq 0 \quad \text{and} \quad \mathbf{f}_k^T(\dot{\mathbf{q}} + \delta\dot{\mathbf{q}}^c) = 0 \tag{9.70}$$

if the contact remains during the compression phase, or

$$\gamma_k^c = 0 \quad \text{and} \quad \mathbf{f}_k^T(\dot{\mathbf{q}} + \delta\dot{\mathbf{q}}^c) > 0 \tag{9.71}$$

if it breaks. $\gamma_k^c$ is the compression impulse magnitude on contact $k$. Combining the constraints on all contacts and reformulating gives

$$\text{diag}(\gamma^c)(\mathbf{F}^T\dot{\mathbf{q}} + \mathbf{F}^T\delta\dot{\mathbf{q}}^c) = \mathbf{0} \tag{9.72}$$

and

$$\gamma^c + \mathbf{F}^T\dot{\mathbf{q}} + \mathbf{F}^T\delta\dot{\mathbf{q}}^c \geq \mathbf{0}. \tag{9.73}$$

The equation of impulsive motion for the contact-free system subject to contact impulses is (from Equation (9.49))

$$\mathbf{I}\,\delta\dot{\mathbf{q}}^c = \mathbf{F}\,\gamma^c. \tag{9.74}$$

Substituting for $\delta\dot{\mathbf{q}}^c$ in Equations (9.72 and 9.73) gives

$$\text{diag}(\gamma^c)(\mathbf{C}\,\gamma^c + \mathbf{F}^T\dot{\mathbf{q}}) = \mathbf{0} \tag{9.75}$$

and

$$(\mathbf{C} + \mathbf{1})\,\gamma^c + \mathbf{F}^T\dot{\mathbf{q}} \geq \mathbf{0}, \tag{9.76}$$

where $\mathbf{C}$ is as defined in Equation (9.52). The only difference between these equations and Equations (9.50 and 9.51) is that $\mathbf{d}$ is replaced by $\mathbf{F}^T \dot{\mathbf{q}}$.

During the decompression phase, we will assume that each contact exerts an impulse $\gamma_k^d$ given by

$$\gamma_k^d = e_k \gamma_k^c + \gamma_k^x , \tag{9.77}$$

where $e_k$ is the coefficient of restitution at contact $k$, and $\gamma_k^x$ is an additional non-negative impulse which is needed to ensure that the contact constraints are not violated. The set of contacts is the same as during the compression phase, and the contact constraints are, for each contact, either

$$\gamma_k^x \geq 0 \qquad \text{and} \qquad \mathbf{f}_k^T ( \dot{\mathbf{q}} + \delta\dot{\mathbf{q}}^c + \delta\dot{\mathbf{q}}^d ) = 0 \tag{9.78}$$

or

$$\gamma_k^x = 0 \qquad \text{and} \qquad \mathbf{f}_k^T ( \dot{\mathbf{q}} + \delta\dot{\mathbf{q}}^c + \delta\dot{\mathbf{q}}^d ) > 0 . \tag{9.79}$$

$\delta\dot{\mathbf{q}}^d$ is the system velocity change during decompression, and is given by

$$\mathbf{I} \, \delta\dot{\mathbf{q}}^d = \mathbf{F} \, \gamma^d \quad . \tag{9.80}$$

Combining the constraints as before, and substituting for $\delta\dot{\mathbf{q}}^d$, results in the following:

$$\mathrm{diag}(\gamma^x) \, ( \mathbf{C} \, \gamma^x + \mathbf{g} ) = \mathbf{0} \tag{9.81}$$

and

$$( \mathbf{C} + \mathbf{1} ) \, \gamma^x + \mathbf{g} \geq \mathbf{0} , \tag{9.82}$$

where

$$\mathbf{g} = \mathbf{F}^T \dot{\mathbf{q}} + \mathbf{C} \, ( \mathbf{1} + \mathrm{diag}(\mathbf{e}) ) \, \gamma^c , \tag{9.83}$$

and $\mathbf{e}$ is the vector of coefficients of restitution. Two special cases are worth noting: if all $e_k$ are zero then there is no decompression phase, and if all $e_k$

are the same then $\gamma^{\mathcal{I}} = \mathbf{0}$. At the end of the impact, those contacts which are broken should be removed from the set of contacts before resuming continuous motion simulation.

# References

[1] Andrews, G. C. and Kesavan, H. K.
Simulation of multibody systems using the vector-network model.
In K. Magnus (editor), *IUTAM Symposium on the Dynamcis of Multibody Systems*. Springer-Verlag, Berlin, Heidelberg, New York, 1978.

[2] Armstrong, W. W.
Recursive solution to the equations of motion of an n link manipulator.
In *Proc. 5th World Congress on the Theory of Machines and Mechanisms, Vol. 2,* pages 1343-1346. Montreal, July, 1979.

[3] Backhouse, C. and Rees Jones, J.
Analogue computer simulation of a robot manipulator.
*I. Mech. E. Journal of Mechanical Engineering Science* 23(3):121-129, 1981.

[4] Ball, R. S.
*A treatise on the theory of screws.*
Cambridge Univ. Press, London, 1900.

[5] Baumgarte, J.
Stabilization of constraints and integrals of motion in dynamical systems.
*Computer Methods in Applied Mechanics and Engineering* 1:1-16, 1972.

[6] Benati, M., Gaglio, S., Morasso, P., Tagliasco, V., Zaccaria, R.
Anthropomorphic robots.
*Biological Cybernetics* 38:125-150, 1980.

[7] Bottema, O. and Roth, B.
*Theoretical kinematics.*
North Holland, Amsterdam, 1979.

[8] Boyse, J. W.
Interference detection among solids and surfaces.
*Comm. ACM* 22(1), 1979.

[9] Brach, R. M.
Moments between impacting rigid bodies.
*Trans ASME Jnl. Mechanical Design* 103(4):812-817, 1981.

[10] Brady, J. M., Hollerbach, J. M., Johnson, T. L., Lozano-Perez, T., and Mason, M. T.
*Robot motion: planning and control.*
MIT Press, Cambridge, MA; London, 1982.

[11] Brand, L.
*Vector and tensor analysis.*
Wiley/Chapman and Hall, New York/London, 1953.

[12] Cameron, S. A.
*Modelling solids in motion.*
PhD thesis, Dept. of Artificial Intelligence, Univ. of Edinburgh, 1984.

[13] Clifford, W. K.
Preliminary sketch of biquaternions.
*Mathematical Papers.*
MacMillan, London, 1882.

[14] de Pennington, A., Bloor, M. S., and Balila, M.
Geometric modelling: a contribution towards intelligent robots.
In *Proc. 13th International Symposium on Industrial Robots, Vol. 1*, pages 35-54. Chicago, April, 1983.

[15] Denavit, J. and Hartenberg, R. S.
A kinematic notation for lower pair mechanisms based on matrices.
*Trans ASME Jnl. Applied Mechanics* 22:215-221, 1955.

[16] Derby, S. J.
*Kinematic elasto-dynamic analysis and computer graphics simulation of general purpose robot manipulators.*
PhD thesis, Rensselaer Polytechnic Institute, 1981.

[17] Dimentberg, F. M.
*The screw calculus and its applications in mechanics.*
Izd. Nauka. NOTE: In Russian. English translation: Foreign Technology Div., Wright-Patterson AFB, Ohio, Part No. AD 680 993, 1968., Moscow, 1965.

[18] Duffy, J.
*Analysis of mechanisms and robot manipulators.*
Edward Arnold, London, 1980.

[19] Featherstone, R.
*A program for simulating robot dynamics.*
Working Paper 116, Dept. of AI, Univ. of Edinburgh, 1982.

[20] Featherstone, R.
*Update on the robot simulator program.*
Working Paper 155, Dept. of AI, Univ. of Edinburgh, 1983.

[21] Featherstone, R.
The calculation of robot dynamics using articulated-body inertias.
*Robotics Research* 2(1):13-30, 1983.

[22] Hollerbach, J. M.
A recursive Lagrangian formulation of maniplator dynamics and a comparative study of dynamics formulation complexity.
*IEEE Trans. Systems, Man and Cybernetics* SMC-10(11):730-736, 1980.

[23] Hollerbach, J. M.
Dynamic scaling of manipulator trajectories.
*Trans. ASME Jnl. Dynamic Systems Measurement and control* 106:102-106, 1984.

[24] Hollerbach, J. M. and Sahar, G.
Wrist-partitioned inverse kinematic accelerations and manipulator dynamics.
*Robotics Research* 2(4):61-76, 1983.

[25] Hooker, W. W.
A set of r dynamical attitutde equations for an arbitrary n-body satellite having r rotational degress of freedom.
*AIAA Journal* 8(2)(7):1205-1207, 1970.

[26] Hooker, W. W. and Margulies, G.
The dynamical attitude equations for an n-body satellite.
*Jnl. Astronautical Sciences* 12(4):123-128, 1965.

[27] Horn, B. K. P. and Raibert, M. H.
*Configuration space control.*
AI Memo 458, MIT AI Lab, 1977.

[28] Hunt, K. H.
*Kinematic geometry of mechanisms.*
Clarendon Press, Oxford, 1978.

[29] Huston, R. L., Passerello, C. E., and Harlow, M. W.
Dynamics of multirigid-body systems.
*Trans. ASME Jnl. Applied Mechanics* 45:889-894, 1978.

[30] Kahn, M. E. .
*The near minimum-time control of open-loop articulated Kinematic chains.*
AI Memo 106, Stanford AI Lab., 1969.

[31] Kane, T. R. and Faessler, H.
Dynamics of robots and manipulators involving closed loops.
In Morecki, et.al. (editor), *Theory and Practice of Robots and Manipulators*, pages 97-106. Kogan Page Ltd., London, 1985.

[32] Kane, T. R. and Levinson, D. A.
The use of Kane's dynamical equations in robotics.
*Robotics Research* 2(3):3-21, 1983.

[33] Khatib, O.
Dynamic control of manipulators in operational space.
In *6th IFToMM Congress on Theory of Machines and Mechanisms*. New Dehli, Dec., 1983.

[34] Kreuzer, E. J. and Schiehlen, W. O.
Equations of motion and equations of stress for robots and manipulators: an application of the NEWEUL Formalism.
In Morecki, et al. (editor), *Theory and Practice of Robots and Manipulators*, pages 79-85. Kogan Page Ltd., London, 1985.

[35] Lang, S.
*Linear algebra.*
Addison-Wesley, Reading, MA, 1972.

[36] Lipkin, H. and Duffy, J.
Analysis of industrial robots via the theory of screws.
In *Proc. 12th International Symposium on Industrial Robots*, pages 359-370. Paris, June, 1982.

[37] Lotstedt, P.
Mechanical systems of rigid bodies subject to unilateral constraints.
*SIAM Jnl. Applied Mathematics* 42(2):281-296, 1982.

[38] Lotstedt, P.
Numerical simulation of time-dependent contact and friction problems in rigid body mechanics.
*SIAM Jnl. Scientific and Statistical Computing* 5(2):370-393, 1984.

[39] Luh, J. Y. S., Walker, M. W., and Paul, R. P. C.
Resolved acceleration control of mechanical manipulators.
*IEEE Trans. Automatic Control* AC-25(3):468-474, 1980.

[40] Luh, J. Y. S., Walker, M. W., and Paul, R. P. C.
On-line computational scheme for mechanical manipulators.
*Trans. ASME Jnl. Dynamic Systems Measurement and Control* 102(2):69-76, 1980.

[41] Mufti, I. H.
*Equations of motion for rigid multibody systems.*
paper DM-3 NRC 22797, National Research Council of Canada, 1983.

[42] Ohwovoriole, M. S.
*An extension of screw theory and its application to the automation of industrial assemblies.*
PhD thesis, Dept. of Computer Science, Stanford Univ., 1980.

[43] Orin, D. E., McGhee, R. B., Vukobratovic, M., and Hartoch, G.
Kinematic and kinetic analysis of open-chain linkages utilising Newton-Euler methods.
*Mathematical Biosciences* 43:107-130, 1979.

[44] Orlandea, N., Calahan, D. A., and Chace, M. A.
A sparsity-oriented approach to the dynamic analysis and design of mechanical systems.
*Trans. ASME Jnl. Engineering for Industry* 99(3):773-784, 1977.

[45] Paul, B.
Analytical dynamics of mechanisms - a computer oriented overview.
*Mechanism and Machine Theory* 10:481-507, 1975.

[46] Paul, R.
*Robot manipulators: mathemactics, programming and control.*
MIT Press, Cambridge, MA; London, 1981.

[47] Renaud, M.
An efficient iterative analylytical procedure for obtaining a robot manipulator dynamic model.
In *Proc. 1st International Symposium on Robotics Research.* Bretton Woods, USA, Aug. 28 - Sept. 2, 1983.

[48] Roberson, R. E.
Constraint stabilisation for rigid bodies: an extension of Baumgarte's method.
In K. Magnus (editor), *IUTAM Symposium on the Dynamics of Multibody Systems.* Springer-Verlag, Berlin, Heidelberg, New York, 1978.

[49] Rooney, J.
On the principle of transference.
In *Proc. 4th World Congress on the Theory of Machines and Mechanisms, Vol. 5,* pages 1089-1094. Newcastle upon Tyne, Sept., 1975.

[50] Rooney, J.
A survey of representations of spatial rotation about a fixed point.
*Environment and Planning B* 4:185-210, 1977.

[51] Rooney, J.
A comparison of representations of general spatial screw displacement.
*Environment and Planning B* 5:45-88, 1978.

[52] Sahar, G. and Hollerbach, J. M.
*Planning of minimum-time trajectories for robot arms.*
AI Memo 804, MIT AI Lab., 1984.

[53] Schwertassek, R. and Roberson, R. E.
A state-space dynamical representation for multibody mechanical systems, Part 1: systems with tree configuration.
*Acta Mechanica* 50:141-161, 1984.

[54] Schwertassek, R. and Roberson, R. E.
A state-space dynamical representation for multibody mechanical systems, Part 2: systems with closed loops.
*Acta Mechanica* 51:15-29, 1984.

[55] Schwertassek, R. and Roberson, R. E.
Computer-aided generation of multibody-system equations.
In Morecki, et al. (editor), *Theory and Practice of Robots and Manipulators*, pages 73-78. Kogan Page Ltd., London, 1985.

[56] Sheth, P. N. and Uicker, Jr., J.J.
A generalised symbolic notation for mechanisms.
*Trans. ASME Jnl. Engineering for Industry* 93:102-112, 1971.

[57] Silver, W.
On the equivalence of Lagrangian and Newton-Euler dynamics for manipulators.
*Robotics Research* 1(2):60-70, 1982.

[58] Smith, D. A.
Reaction force analysis in generalised machine systems.
*Trans. ASME Jnl. Engineering for Industry* 95:617-623, 1973.

[59] Smith, D. A., Chace, M. A., and Rubens, A. C.
The automatic generation of a mathematical model for machinery systems.
*Trans. ASME Jnl. Engineering for Industry* 95:629-635, 1973.

[60] Stepanenko, Y., and Vukobratovic, M.
Dynamics of articulated open chain active mechanisms.
*Mathematical Biosciences* 28:137-170, 1976.

[61] Swartz, N. M.
Arm dynamics simulation.
*Jnl. Robotic Systems* 1(1):83-100, 1984.

[62] Thomas, M. and Tesar, D.
Dynamic modelling of serial manipulator arms.
*Trans ASME Jnl. Dynamic Systems Measurement and Control* 104(1):218-228, 1982.

[63] Tilove, R. B.
*Exploiting spatial and structural locality in geometric modelling.*
PhD thesis, College of Engineering and Applied Science, Univ. of Rochester, 1981.

[64] Turney, J. L., Mudge, T. N., and Lee, C. S. G.
*Equivalence of two formulations for robot arm dynamics.*
Technical Report, Robot Sys. Div. (CRIM), Univ. of Michigan, 1980.

[65] Udupa, S. M.
Collision detection and avoidance in computer-controlled manipulators.
In *Proc. 5th International Joint Conference on Artificial Intelligence, Vol. 2.* Boston, 1977.

[66] Uicker, Jr., J. J.
*On the dynamic analysis of spatial linkages using 4x4 matrices.*
PhD thesis, Northwestern University, Evanston, IL, 1965.

[67] Uicker, Jr., J. J.
Dynamic force analysis of spatial linkages.
*Trans. ASME Jnl. Applied Mechanics* 34:418-424, 1967.

[68] Vereshchagin, A. F.
Computer simulation of the dynamics of complicated mechanisms of robot manipulators.
*Engineering Cybernetics* (6):65-70, 1974.

[69] von Mises, R.
Motorrechnung, ein neues Hilfsmittel der Mechanik [Motor Calculus: a new tool for mechanics].
*Zeitschrift fur Angewandte Mathematik und Mechanik* 4(2):155-181, 1924.

[70] von Mises, R.
Anwendungen der Motorrechnung [Applications of Motor Calculus].
*Zeitschrift fur Angewandte Mathematik und Mechanik* 4(3):193-213, 1924.

[71] Vukobratovic, M.
Dynamics of active articulated mechanisms and synthesis of artificial motion.
*Mechanism and Machine Theory* 13:1-56, 1978.

[72] Vukobratovic, M. and Kircanski, N.
Implementation of highly efficient analytical robot models on microcomputers.
In Morecki, et al. (editor), *Theory and Practice of Robots and Manipulators*, pages 65-71. Kogan Page Ltd., London, 1985.

[73] Vukobratovic, M. and Potkonjak, V.
Contribution of the forming of computer models for automatic modelling of spatial mechanisms motions.
*Mechanism and Machine Theory* 14:179-188, 1979.

[74] Walker, M. W. and Orin, D. E.
Efficient dynamic computer simulation of robotic mechanisms.
*Trans. ASME Jnl. Dynamic Systems Measurement and Control* 104:205-211, 1982.

[75] Waters, R. C.
*Mechanical arm control.*
AI Memo 549, MIT AI Lab, 1979.

[76] Wittenburg, J.
*Dynamics of systems of rigid bodies.*
B. G. Teubner, Stuttgart, 1977.

[77] Woo, L. S. and Freudenstein, F.
Application of line geometry to theoretical kinematics and the kinematic analysis of mechanical systems.
*Journal of Mechanisms* 5:417-460, 1970.

[78] Woo, L. S. and Freudenstein, F.
Dynamic analysis of mechanisms using screw coordinates.
*Trans. ASME Jnl. Engineering for Industry* 93:273-276, 1971.

[79] Yang, A. T.
Inertia force analysis of spatial mechanisms.
*Trans. ASME Jnl. Engineering for Industry* 93:27-33, 1971.

[80] Yang, A. T. and Freudenstein, F.
Application of dual number quaternion algebra to the analysis of spatial mechanisms.
*Trans. ASME Jnl. Applied Mechanics* 31:300-308, 1964.

[81] Yuan, M. S. C. and Freudenstein, F.
Kinematic analysis of spatial mechanisms by means of screw coordinates. Part 1 - Screw Coordinates.
*Trans. ASME Jnl. Engineering for Industry* 93:61-66, 1971.

[82] Yuan, M. S. C., Freudenstein, F., and Woo, L. S.
Kinematic analysis of spatial mechanisms by means of screw coordinates. Part 2 - analysis of spatial mechanisms.
*Trans. ASME Jnl. Engineering for Industry* 93:67-73, 1971.

# Index